한반도, 정전에서 평화로

The Korean Peninsula, Armistice to Peace

이 도서의 국립중앙도서관 출판예정도서목록(CIP)은 서지정보유통지원시스템 홈페이지(http://seoji.nl.go.kr)와 국가자료종합목록시스템(http://www.nl.go.kr/kolisnet)에서 이용하실 수 있습니다.
(CIP제어번호 : CIP2019017835)

한반도, 정전에서 평화로

The Korean Peninsula, Armistice to Peace

신상범 편저

한누리미디어

Prologue

As the Senior Member of the United Nations Command Military Armistice Commission (UNCMAC), I wanted to help non-military readers better understand the changes and transformation of the security environment on the Korean Peninsula by writing a simple and easy-to-follow book that explains such topics as the Korean War, the establishment of United Nations Command (UNC), the implementation of the Armistice Agreement, the 2018 inter-Korean agreements between the two leaders, and the Comprehensive Military Agreement (CMA) for the implementation of the Panmunjom Declaration.

The year 2018 marked a new beginning of peace and prosperity on the Korean Peninsula. To reap the fruits from the seeds of change we planted last year, we must continue to lay the groundwork for a Korean Peninsula that is free of war and enjoys irreversible peace and prosperity by fully understanding the past and present of our security environment. This includes negotiating and signing the Armistice Agreement, maintaining the Armistice regime on the Korean Peninsula, and implementing the CMA.

My deepest gratitude goes to our comrades who came before us, and the UNC with its Sending States that have provided trust and support

책머리에

필자는 유엔군사령부 군사정전위원회 수석대표로서 한반도 안보상황의 변천과정, 즉 6.25전쟁과 유엔군사령부 창설 그리고 정전협정 이행, 2018년 남북정상이 만나서 합의한 내용과 판문점 선언 이행을 위한 군사분야합의서 등을 일반국민들이 쉽게 이해하는 데 도움을 주고자 관련 내용을 보고 드리는 마음으로 정리하였습니다.

• 군사정전위원회 : 정정협정시 공산군측과 유엔군측은 정전협정이행을 관리할 기구로 양측 각 5명씩 모두 10명으로 군사정전위원회를 구성하였다. 하지만 공산군측은 1994년도에 철수하였고, 현재는 유엔군측 5명이 유지하고 있음.

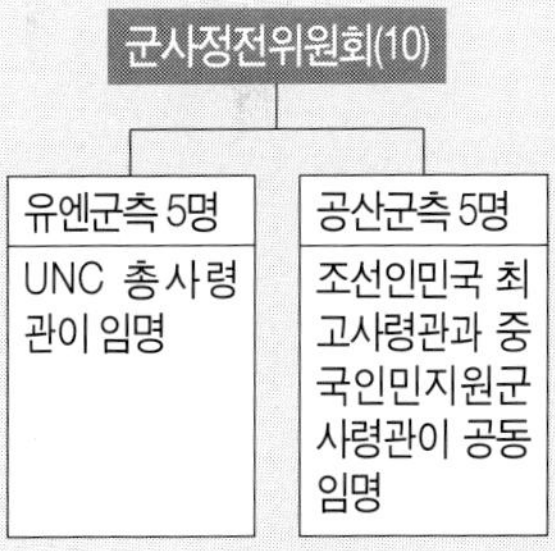

2018년은 한반도의 평화와 번영을 알리는 새로운 시작이었습니다. 이러한 변화가 성공의 열매를 얻기 위해서는 한국전쟁과 유엔군사령부 설치, 휴전회담과 정전협정 체결, 한반도 정전체제 유지와 9.19 군사분야 합의 등 안보상황 변화에 대한 충분한 이해를 바탕으로 전쟁이 없는 그리고 불가역적인 한반도의 평화와 번영을 위한 준비를 만들어 나가야 하겠습니다.

무엇보다도 정전협정이 오늘날까지 잘 이행되어 대한민국의 평화와 번영을 이룩할 수 있도록 신뢰와 버팀목

for the Republic of Korea to achieve peace and prosperity by ensuring the implementation of the Armistice Agreement. Most of all, I am grateful for everyone who has dedicated their lives and youth for the peace and prosperity of this country.

It is my hope that this small book will serve as a helpful guide, even if only a little, for readers to understand armistice and peace. Lastly, I would also like to thank Writer Jeon Jeong-hee for her great assistance in writing this book, all those who helped me with the publication and English translation of this book, and my family. I wish my motherland, the Republic of Korea, which will forever protected by with all our men and women in uniform working so tirelessly to defend this country even at this moment, everlasting peace and prosperity.

At the conference room where the Geneva Conference was held in 1954
February 9, 2019

MG **Shin Sang-bum**, Senior Member, UNCMAC

이 되어주신 국군장병과 선배전우 여러분, 유엔군사령부 및 전력지원제공국, 특히 자신의 삶과 젊음을 대한민국의 평화와 번영을 위해 희생하신 모든 분들께 깊이 감사드립니다.

바라기는 이 작은 책자가 정전과 평화를 이해하는 데 조금이나마 도움이 되기를 소망합니다. 끝으로 출판에 도움을 주신 전정희 작가님, 출판과 번역에 관계하신 분들, 한미동맹에 힘쓰시는 우현의 회장님, 사랑하는 나의 가족과 아들 유찬 · 재원에게 감사함을 전합니다. 아울러 오늘도 불철주야 나라를 지키기 위해 헌신하고 애쓰는 국군장병과 함께 영원히 빛날 조국 대한민국의 평화와 번영을 기원합니다.

1954년 정치회담이 열렸던 제네바 회의장에서
2019년 2월 9일

유엔군사령부 군사정전위원회
수석대표 육군소장 **신 상 범**

회의장 전경

A review of 'The Korean Peninsula, Armistice to Peace'

I think it was The spring of 2010 when I volunteered to work at Dorasan Peace Park, located in Jangdan-myeon, Paju City, Gyeonggi Province. Our work as volunteers was to escort visitors, help with the landscaping, and maintain the eco-friendly ponds at the park.

After the groundbreaking on May 12, 2006, Dorasan Peace Park opened to the public on September 10, 2008. Since it was located north of the Civilian Control Line, I had to go through an identification(ID) check for access before I could board a train from Imjin Station. I volunteered to work at the park; thinking how amazing it would be if one day people from the South and the North could walk together in a beautiful park, and I wanted to be part of it. But my rewarding experience as a volunteer had to stop after the second visit. On March 26, 2010, the ROK ship Cheonan was sunk. In the waters off of Baengnyeong-Island, PCC-772 Cheonan, a ROK Navy's patrol ship, was sunk by a torpedo, resulting in 40 sailors killed and six missing. The relationship between the South and the North quickly turned sour so much so that it seemed as if war was imminent. As a result, access to the Peace Park was restricted and the volunteer work was also suspended. On my first day, I sowed flower seeds in a flowerbed and stripped trees of the straw wraps that kept them warm all winter. I

《한반도, 정전에서 평화로》를 읽고

도라산 평화공원

2010년 봄으로 기억된다. 나는 경기도 파주시 장단면에 위치한 도라산 평화공원 자원활동가(지킴이)로 봉사를 한 적이 있었다. 지킴이들은 도라산 평화공원의 안내와 홍보, 공원의 조경과 생태연못 등을 가꾸는 일을 맡아 했다.

도라산 평화공원은 2006년 5월 12일에 착공하여 2008년 9월 10일에 개장, 일반인들에게 개방되었다. 이곳은 민간인통제구역에 위치하고 있어 소정의 확인절차를 거친 후 임진강역에서 기차를 타고 입장하였다. 이 아름다운 공원을 잘 가꾸어 남과 북이 하나가 되어 거닐 수 있다면 얼마나 좋을까 하는 생각으로 봉사를 신청했고 일도 즐거웠다. 그러나 나는 이 뜻 깊은 봉사를 겨우 두 번밖에 하지 못했다. 그해 3월 26일 천안함 피격사건이 발생했기 때문이었다. 백령도 근처 해상에서 대한민국 해군의 초계함인 PCC-772 천안함이 뇌격되어 침몰하였는데 대한민국 해군 장병 40명이 사망했으며 6명이 실종되었다. 이 일로 남과 북의 관계는 급랭하여 금방이라도 전쟁이 일어날 듯 분위기가 흉흉했다. 이후 일반인에게 개방되었던 평화공원은 다시 민간인 통제구역이 되었고 지킴이 활동 역시 잠정적으로 중단되었다. 처음 봉사를 나가던 날 나는 화단에 꽃씨를 심었고 겨우내 나무둥치

remember my heart warming up at the thought that the flower seedlings would soon blossom and the park would turn beautiful when spring came. It was such a sad ending to such a heartwarming thought.

However, a wind of peace is blowing again on the Korean Peninsula. A window of reconciliation between the two Koreas opened up on June 24, 2017, when President Moon Jae-in invited North Korea to the PyeongChang Winter Olympic Games at his congratulatory remarks at the World Taekwondo Federation Championship in Muju. Through this window, a series of events were held including a Unified Women's Ice Hockey Team game and a Joint Art Performance wishing for South-North peace and cooperation titled "We are One" in Pyongyang. With such efforts, the year 2018 became a very meaningful year in the history of the Korean people.

On April 27, 2018, the leaders of South and North Korea met in Panmunjom and proclaimed the "Panmunjom Declaration for Peace, Prosperity and Unification of the Korean Peninsula" and the entire world was watching as the two leaders stepping over the Military Demarcation Line(MDL)* separating North and South Korea in the Joint Security Area(JSA)* with big smiles on their faces and their hands tightly locked together. What a moment it was.

The Korean War broke out on June 25, 1950, and the Armistice Agreement was signed on July 27, 1953. If you look up in the dictionary, the word "Armistice" is defined as "a temporary suspension of combat as agreed by the two belligerent sides engaged in military hostilities." This means, literally, that the war is not over, but

를 감싸고 있던 '잠복소' 를 벗겨내 불에 태웠다. 날씨가 따뜻해지면 우리가 심은 꽃씨들이 아름답게 피어나 공원을 빛내줄 생각을 하니 마음이 훈훈했었는데 참으로 안타까운 결말이었다.

그런데 다시 한반도에 평화의 바람이 불어오고 있다. 지난 2017년 6월 24일 문재인 대통령이 무주 WTF세계태권도선수권대회 축사에서 북한의 평창 동계 올림픽 참가를 제안하며 시작한 남북 화해의 장이 남북은 여자 아이스하키 남북단일팀 경기, 남북평화협력기원 예술단 합동 공연 '우리는 하나' (평양) 등의 행사를 이어갔다. 이후 2018년은 우리 역사상 매우 의미 있는 한 해였다.

4월 27일, 판문점에서 남북한 지도자가 만나 '한반도의 평화와 번영, 통일을 위한 판문점 선언' 을 하였으며 두 정상이 손을 마주잡고 공동경비구역(JSA)의 턱을 환하게 웃으며 넘는 사진이 전 세계적으로 전파를 탔다. 참으로 감격스러운 모습이었다.

• MDL : Military Demarcation Line
• JSA : Joint Security Area

1950년 6월 25일 6.25한국전쟁이 발발하였고 1953년 7월 27일 정전협정이 맺어졌다. 정전(停戰)의 사전적 의미는 '군사 교전중에 있는 양방이 합의에 따라 일시적으로 전투를 중단하는 일' 을 말한다. 이는 말 그대로 전쟁이 끝난 것이 아니라 전쟁 상태를 잠시 멈추는 휴전을 뜻하

is just suspended. This implies that fighting can resume at any moment. In other words, although the war came to a halt, we have lived with a risk of war hanging over our heads for decades. However, now that this state of peace, or non-peace, has lasted for 66 years since the signing of the Armistice Agreement I wonder how many people living in South Korea actually understand this reality.

The book 'The Korean Peninsula, Armistice to Peace' is a very enlightening work and helped me learn about the things I didn't understand. For instance, with regard to the Military Demarcation Line (MDL), I learned that the Communist side demanded the 38th parallel be the MDL and each side be withdrawn 10km from the line to establish a Demilitarized Zone, whereas the UNC side proposed a 20 mile-wide DZ. After numerous talks and negotiations, the two sides agreed to have the MDL be the contact points between the two sides at that time and the Demilitarized Zone(DMZ)* be 4km wide at the 28th meeting. I also learned from the book that the 155-mile long MDL is not a fence line, but an invisible line marked by a total of 1,292 markers from Jangdan Peninsula at the estuary of the Imjin River all the way out to the East Coast.

The author stated, "his heart was filled up with emotions" as he was looking at a site where a guard post(GP)* had been demolished in the DMZ with members of the United Nations Command Military Armistice Commission(UNCMAC)* and its Secretariat. There were tears of pride, the author wrote, running down his heart as he was standing at the very place that his comrades before him had defended with their lives through ambush operations in long winter nights and reconnaissance and patrol operations in the mine-filled DMZ for

며 휴전이란 언제 다시 전쟁으로 이어질지 알 수 없음을 암시하고 있다. 즉 전쟁은 정지되었지만 우리는 언제든 전쟁이 재개될 위험을 안고 지금까지 살아온 것이다. 그러나 정전이 된 지 66년이 지나도록 평화 아닌 평화가 유지되고 있으니 대한민국에 살고 있으면서 이 사실을 인지하고 사는 사람은 과연 얼마나 될까 싶은 의문이 든다.

《한반도, 정전에서 평화로》 책을 읽고 평소 잘 알지 못했던 사실들을 일깨우게 되었다. 군사분계선 설정과정이 공산군측은 38도선을 군사분계선으로 설정하고 그 38도선을 중심으로 쌍방 10km의 비무장지대를 설치할 것을 주장한 반면 유엔군측은 20마일의 비무장지대를 제시하였고 결국 수많은 협의 끝에 제28차 본회의에서 현 전선의 접촉선을 지상군사분계선으로 설정하고 비무장지대는 넓이를 4km로 하자는 데 양측은 합의하였다는 내용이다. 뿐만 아니라 임진강 하구 장단반도에서 시작하여 동해안까지 155마일에 이르는 군사분계선이 철책선이 아니고 1,292개의 표식작업을 했다는 내용도 알게 되었다.

저자는 그 DMZ(비무장지대) 내 GP(감시초소) 철수 현장을 군사정전위원회 위원 및 비서처 인원과 함께 방문하고 GP를 무너뜨린 현장을 보면서 만감이 교차함을 느꼈다고 썼다. 그리고 이 비무장지대내에 있는 감시초소들은 오랜 기간 수많은 선배 전우 여러분들이 혹한의 추위 속에서 동지섣달 기나긴 밤에 매복 작전으로 지켜왔고, 수많은 지뢰가 흩어져 있는 비무장지대에 목숨을 걸고 수색정찰하면서 지켜왔던 지역이기 때문에 가슴속에

- DMZ : Demilitarized Zone
- GP : guard post
- UNCMAC : United Nations Command Military Armistice Commission

decades. As I was reading his book, my heart was also filled up with pride and gratitude for all those men and women in uniform. I realized that we owe it to them and their service of tirelessly defending this country for the peace that we enjoy today.

There are many reasons why Koreas should be reunited, but the most important one is probably so that we can live peacefully with tensions reduced between the two Koreas. The risk of war on the Korean Peninsula will be fundamentally eliminated if the two Koreas are unified. The North and the South will no longer have to compete for war, which will eliminate the so-called "cost of division." In addition, a unified Korea will have a logistical gateway to the Eurasian continent. By becoming a transportation and logistical hub connecting Eurasia and the Pacific Ocean, a unified Korea will be put right in the center of the global economy. Goods produced from and imported to South Korea can be transported to North Korea, China, Mongolia, Russia and Europe through railways. Also, trains from a unified Korea can be connected all the way to Paris, Rome and London. Such a logistical network will give Korea a huge competitive edge.

Furthermore, Korea will become a confident unified nation with higher credit ratings as the so called "Korea discount," caused by the confrontations between the North and the South, will be removed. If North Korea keeps launching missiles and heightening tangible threats, foreign and domestic investors will become anxious and unwilling to invest in Korea, which will drive down Korea's credit ratings as well as its national image. Unification, on the contrary, will eliminate all these risks and push up the country's credit ratings.

는 뜨거운 눈물이 흘러내렸다고 적고 있다. 이 글을 읽으면서 나 역시 불철주야 나라를 지키는 군인들로 인해 우리의 평화가 지켜지고 있다는 생각에 그들이 한없이 자랑스럽고 고마웠다.

한반도가 통일을 해야 하는 이유는 여러 가지가 있지만 그 중 가장 중요한 것은 남북한의 긴장완화로 평화로운 삶을 살기 위해서일 것이다. 통일이 되면 한반도에서의 전쟁 가능성을 근원적으로 해결할 수 있다. 따라서 남과 북이 전쟁을 위해 불필요한 경쟁을 하지 않게 되어 소위 말하는 분단비용을 대폭 줄일 수 있다. 또 대륙을 잇는 물류통로가 확보된다. 통일이 되면 한반도를 유라시아 대륙과 태평양을 연결하는 교통과 물류의 핵심으로 만들어 세계 경제의 중심으로 성장할 수 있다. 남한에서 생산한 물건과 수입한 물건이 철도를 통해 북한, 중국, 몽고, 러시아, 유럽까지 연결될 수 있다. 또 통일한국의 열차가 파리, 로마, 런던까지 연결이 가능하다. 즉 물류 시스템이 갖춰지므로 국가 경쟁력에서 앞서게 되는 것이다.

그리고 국가신용등급 상승으로 당당한 통일국가로서 역할을 할 수 있다. 남북이 대치되어 있는 현실로 인해 발생하는 '코리아디스카운트' 라는 부정적 요소가 제거되고 국가 신용등급이 올라갈 것이다. 작금처럼 북한이 미사일을 계속 발사하고 실제 위협요소를 키우면 국외의 투자자들은 불안하기 마련이다. 투자심리가 위축되고 신용등급이 하락하는 상황을 맞을 수도 있다. 국가 브랜드 역시 추락할 수 있다. 그러나 통일이 되면 이러한 불안한 상황이 없어지기 때문에 국가신용등급 상향을

Goldman Sachs, a global investment firm, predicted that the economy of a unified Korea would outgrow that of Japan, Germany and France within 30 to 40 years thanks to the synergy between South Korea's technology and capital and North Korea's resources and manpower. The investment firm also said that the unified Korea would become the seventh largest economy in the world with a per capita income of USD 80,000 by 2050. I believe these predictions are predicated on such factors as increased trade and reduced logistical costs through railways and highly lucrative tourism opportunities connecting Mt. Seorak and Mt. Kumgang.

Since the Armistice agreement in 1953, the two Koreas have held numerous discussions that resulted in the 1992 Inter-Korean Basic Agreement, the June 15 Summit Declaration, the October 10 Joint Declaration, and the 2018 Inter-Korean Summit and Comprehensive Military Agreement, and the efforts still continue. Finally, President Moon Jae-in and Chairman Kim Jung-un planted a tree together for peace and prosperity on the Korean Peninsula at Panmunjom in April 2018. Also, expectations are mounting that inter-Korean economic cooperation may resume contingent on progress with North Korea's denuclearization process with more summits between the South and the North and between the United States (U.S.) and the North scheduled in 2019.

It was our unwavering faith in liberal democracy and our national power to back it up that allowed us to defend our security even in the face of North Korea's constant provocations and aggression. The

기대할 수 있는 것이다.

세계적인 투자회사 골드만삭스는 한반도가 통일될 경우 남한의 기술과 자본, 그리고 북한의 자원과 인력의 시너지 효과로 통일한국의 경제 규모는 30~40년내에 일본과 독일, 프랑스를 추월할 것으로 전망했다. 그리고 통일이 된다면 2050년경에는 국민소득 8만 달러에 이르는 세계 7대 강국이 될 수 있을 것이라고 예상했다. 아마도 이러한 결과는 철도를 통한 무역과 물류비용 절감, 설악산과 금강산을 연결하는 관광 한국의 비전이 엄청난 수익을 창출할 것임을 예견한 결과가 아닐까 생각된다.

1953년 정전협정 이후 우리는 1992년 남북기본합의서, 2000년 6.15정상선언, 2007년 10.4공동선언, 그리고 2018년 남북정상회담 및 군사기본합의서 등 수없이 많은 협의를 해 왔으며 지금도 불철주야 노력하고 있다. 그리고 마침내 2018년 4월 판문점에서 문재인 대통령과 김정은 국무위원장은 한반도의 평화와 번영을 위하여 한 그루의 나무를 심었다. 또 2019년에는 남북정상회담, 북미정상회담 등이 개최되어 북한 비핵화 협상 진전에 따라 남북경협이 재개될 수 있을 것으로 기대를 모으고 있다.

그러나 이처럼 북한의 끊임없는 도발과 침략에도 남한이 안보를 지킬 수 있었던 것은 자유민주주의에 대한 확고한 신뢰와 이를 뒷받침해 온 국력이었다. 국력은 우리

author pointed out that our national power was not just made up of our own, but also of the ROK-U.S. alliance that has been with us since 1953 and the United Nations Command with its Sending States that is committed to standing shoulder to shoulder with us again should there ever be another conflict on the Korean Peninsula. I could not agree more with the author when he said that we were able to defend everything we have thanks to those men and women in uniform, defending this country faithfully and quietly, and the Korean people who fulfilled their citizenry duties for national defense in many different ways. May this small booklet be a great contribution to peace and prosperity on the Korean Peninsula.

Let there be no more war on the Korean Peninsula··· February 2019

Foreword by Novelist **Jeon Jeong-hee**

의 능력뿐만 아니라 지난 1953년 이후 함께한 한미동맹과 한반도에 전쟁이 다시 일어나면 언제든지 와서 함께 싸우고자 다짐하고 있는 전력제공국 즉, 유엔군의 힘이었다고 이 글을 쓴 저자는 말하고 있다. 또 저자의 말대로 이 모든 것을 지켜낸 힘은 오늘도 묵묵히 국가를 지키는 군인과 다양하게 국방의 의무를 수행해 온 대한민국 국민이 바로 주인공이라는 데 격한 공감을 느낀다. 모쪼록 이 작은 책자가 한반도의 평화와 번영을 이루는 데 큰 기여를 했으면 좋겠다.

2019년 2월,

한반도에 더 이상 전쟁이 없기를 바라며…

소설가 **전 정 희**

소설가 **전정희**

- 국제 PEN 한국본부 회원
- 한국문인협회 이사
- 한국소설가협회 이사
- 한국소설창작연구회 이사
- 2018년 대한민국 인권상 대상 수상

Table of Contents

I. Armistice

차례

제1장 정전

Table of Contents

II. Peace

제2장 평화

I

정전
Armistice

Chapter 1

1. On Armistice

The Korean War, which broke out on June 25, 1950 with an unprovoked invasion of South Korea by North Korea, caused numerous human casualties and property losses, and lasted for three years until an The Armistice was signed. To conclude this devastating war peacefully, the Korean Armistice Agreement was signed on July 27, 1953 by General Mark W. Clark, Commander-in-Chief of the United Nations Command (UNC), Marshal Kim Il-sung, Democratic People's Republic of Korea Supreme Commander of the Korean People's Army (KPA) and Peng Teh-huai, Commander of Chinese People's Volunteers (CPV).

Since its signing, the Armistice Agreement has been the standard for maintaining military order on the Korean Peninsula for both South and North Korea. The Military Demarcation Line (MDL) and the Demilitarized Zone (DMZ) agreed upon in the Armistice Agreement have physically prevented the South and the North from committing major hostilities against each other. The Military Armistice Commission(MAC) and the Neutral Nations Supervisory Commission (NNSC), which were also formed in 1953, have greatly contributed to the economic and national growth of the Republic of Korea by ensuring a stable security environment, despite numerous challenges.

Chapter 1

1. 정전에 대하여

1950년 6월 25일 시작된 6.25한국전쟁은 1953년 7월 27일 정전협정이 체결될 때까지 약 3년간 계속되었고 남 · 북한에 수많은 인명피해와 재산의 손실이 발생하였습니다. 이 참혹한 전쟁을 평화적으로 마무리하기 위하여 국제연합군 총사령관 미국 육군대장 마크 더블유 클라크(Mark W. Clark), 조선인민군 최고사령관 조선민주주의 인민공화국 원수 김일성, 중국인민지원군 사령관 팽더화이(彭德懷)는 1953년 7월 27일 정전협정에 서명을 하였습니다.

정전협정서명
Signing the Armistice Agreement

정전협정은 그 서명 이후 지금까지 한반도의 남한과 북한의 군사질서 유지의 기준이 되어 왔습니다. 당시 합의된 군사분계선(MDL : Military Demarcation Line)과 비무장지대(DMZ : Demilitarized Zone)는 지리적으로 남한과 북한이 서로에게 적대행위를 하지 못하도록 하였고, 당시에 구성된 군사정전위원회와 중립국감독위원회는 많은 어려움은 있었으나 우리나라가 안정적인 안보환경에서 경제발전과 국력을 신장하는 데 크게 기여해 왔습니다.

In March 2018, during a ROK and U.S. combined exercise, a U.S. Major who was augmenting the U.S. Forces Korea from the mainland U.S. asked me, "Do you know who BTS is?" I was surprised by the question because I didn't expect him to be familiar with any South Korean entertainers. In fact, I had no idea how huge BTS was back then. Soon after, BTS gave a speech at the United Nations (UN) Headquarters in New York with the theme of "Love your life," which was a message to young people around the world. I was truly proud of those young Koreans who were promoting their country with music in the global stage.

The Korean people, who have lived on the Korean Peninsula for thousands of years, were liberated from Japan on August 15, 1945, but divided into the democratic South and the communist North across the 38th parallel by the Allied Forces sent to disarm the occupying Japanese troops (The Korean War and UN Forces; p. 467). Soon after, the Korean War broke out on June 25, 1950, with the Communist forces of the North invading the South. After much turmoil, the Armistice Agreement was signed on July 27, 1953 putting a truce on the devastating three year-long war. The war turned everything into ashes and the Korean people couldn't even feed themselves without the aid and support from foreign countries, including the U.S.

66 years later, in 2019, the Republic of Korea has become a mature democracy with remarkable economic success. Also, in sport and cultural arenas, South Korean woman golfers have excelled in the global stage and its musicians are leading global cultural trends. In short, the Republic of Korea has evolved into a whole new and different country.

2018년 3월 한 · 미 연합군사 훈련을 하고 있던 중 미국 본토에서 증원되어 우리나라에 와서 훈련에 참여하고 있던 미군 소령이 본인(저자)에게 "BTS(방탄소년단)를 아느냐?"는 질문을 해 왔습니다. 이 질문을 듣고 그가 우리나라 연예인을 알고 있다는 사실에 깜짝 놀랐습니다. 사실 그때까지만 해도 본인은 방탄소년단의 위상을 잘 알지 못했습니다. 그로부터 얼마 후 BTS(방탄소년단)는 미국 뉴욕에 있는 UN본부에서 지구촌 젊은이들에게 '자신의 삶을 사랑하자' 라는 내용으로 연설을 하였습니다. 음악으로 우리나라의 위상을 세계에 널리 알리는 우리 청년들이 참으로 자랑스러웠습니다.

한반도에서 수천 년의 삶을 살아온 우리 민족은 1945년 8월 15일 일제의 억압으로부터 독립을 하였으나, 일본군의 무장해제를 위해 연합군에 의해 38도선을 기준으로 자유민주주의 체제의 남한과 공산주의 체제의 북한으로 나누어졌습니다.* 그리고 얼마 후인 1950년 6월 25일 공산국가인 북한군의 남한침략으로 6.25한국전쟁이 시작되었고 천신만고 끝에 1953년 7월 27일 정전협정으로 휴전을 얻었으나 약 3년간의 치열한 전쟁으로 모든 것은 잿더미로 변하면서 미국을 포함한 외국의 원조와 지원 없이는 먹고 사는 것조차 힘들었습니다.

그로부터 66년이 지난 2019년 대한민국은 자유민주주의 체제를 정착시켰고 성공적인 경제 성장을 달성하였습니다. 뿐만 아니라, 여자 프로골퍼들의 세계적 수준에서의 활동, 문화를 선도하는 아이돌 그룹 등의 활약 등 과거와 비교를 할 수 없을 정도로 성장하였습니다.

비무장지대내 판문점
Panmunjom in the Demilitarized Zone

* 손규석 · 조성훈 · 김상원, 『6.25 전쟁과 유엔군』(서울 : 국방부 군사편찬연구소, 2015), 467쪽

〈Korean War and Armistice Agreement〉

In response to the communist invasion of South Korea, an illegalact of aggression, the UN Security Council in New York adopted a series of UN resolutions; No. 82 demanding the cessation of act of war and No. 84 authorizing the creation of a unified command. Based on this resolution, General MacArthur was appointed as the Commander of UNC and given the operational control of the ROK Army, Navy and Air Force based on then President Rhee's letter dated July 15, 1950.

After invading the South on June 25, 1950, the North Korean Army quickly advanced south, and South Korea, which had not prepared for war, lost Seoul in just three days. Some U.S. troops arrived in Korea in July, but were unable to repel the North Korean forces armed with Soviet tanks. South Koreans continued to retreat further south and the UNC forces fought fierce battles along the Nakdong Perimeter to defend Daegu and Busan.

In September 1950, about three months after the war began, GEN MacArthur, the Commander of UNC, turned the war around with a successful amphibious operation in Incheon and began advancing northward. The UN forces quickly secured Seoul, passed the 38th parallel, seized Pyongyang, and made it to a point where unification seemed imminent. However, with the Chinese intervention in October 1950, the UN forces had to retreat and the two sides repeated a cycle of retreats and counterattacks, drawing the war out. Against this backdrop, UN member states began calling for a peaceful settlement of the war through diplomatic negotiations and there was a growing

〈6.25한국전쟁과 정전협정〉

공산국가인 북한의 남한에 대한 불법침략에 대하여, 유엔 안전보장이사회는 전쟁행위 중지에 관한 유엔 결의안 82호, 국제연합통합군사령부 설치에 관한 유엔 결의안 84호를 통과시켰습니다. 이를 근거로 하여 맥아더(Douglas MacArthur) 장군이 유엔군 총사령관으로 임명되었으며, 1950년 7월 15일 대한민국 이승만 대통령이 맥아더 장군에게 보낸 서한에 의거 대한민국 육 · 해 · 공군의 작전 지휘권은 유엔군 총사령관에게 이양되었습니다.

이승만 대통령과 맥아더 장군

1950년 6월 25일 남한을 침략한 이후 북한군은 신속하게 남쪽으로 진격하였고 전쟁준비를 하지 않았던 남한은 3일 만에 서울을 잃었으며 얼마 후 7월경에 미군이 일부 참전을 하였으나 소련제 탱크를 앞세운 북한군의 전투력을 방어할 수 없었습니다. 남한 국민들은 남쪽으로, 남쪽으로 후퇴를 거듭하고 대구와 부산만을 남겨둔 유엔군은 낙동강을 중심으로 방어 전선을 펼쳤으며 처절한 전투는 계속되었습니다.

전쟁이 시작되고 약 3개월이 지난 1950년 9월경 맥아더 유엔군 사령관은 인천상륙작전을 성공리에 수행하여 순식간에 전쟁의 주도권을 확보하면서 서울을 회복하였고 북쪽으로, 북쪽으로 계속 전진하여 38선을 지나고 평양을 지나 통일을 목전에 두게 되었습니다. 그러나 1950년 10월경 중공군이 한국전쟁에 개입함에 따라 유엔군은 또다시 남쪽으로 후퇴와 반격을 반복하게 되었습니다. 유엔군이 북진할 때와는 달리 전쟁은 확전되었고 장기전의 양상을 띠게 되었습니다. 이러한 상황에서 유엔 회원국들은 외교적 협상을 통한 평화적 해결을 촉구하

voice for a peaceful truce in the U.S. as well (The Korean Peninsula Armistice Regime, p. 11).

On June 23, 1951, about a year after the outbreak of the Korean War, Soviet Foreign Minister Jacob Malik said in a UN broadcasting program titled 'The Price of Peace', that the Communist side was willing to negotiate the end of the Korean War. On June 30, 1951, GEN Mathew B. Ridgeway, the Commander of UNC, proposed a truce negotiation through a radio program called 'Jutlandia'(JSA Panmunjom, p. 269).

On July 1, 1951, the Communist forces accepted the UNC Commander's offer, and proposed to hold a truce negotiation on July 10 in Kaesong. The two sides held a liaison officer's meeting on July 6 and began a plenary session to discuss the terms of a truce on July 10.

UNC and the Communist forces held their first armistice meeting on July 10, 1951, to find a peaceful end to the Korean War, but for nearly two years until July 27, 1953, fierce battles took place on one side with armistice negotiations on the other.

The two sides held military discussions at the commander level on the following agenda: establishing a DMZ and a MDL to cease hostilities, stopping combat, establishing measures to ensure implementation of the agreement, discussing Prisoners of War (POW) issues, and convening a political meeting on withdrawal of foreign troops and a peaceful resolution to the Korean Peninsula issue (The Korean War and UN Forces, p. 462). However, even today, a peaceful settlement of the Korean War through political discussions has yet to be achieved.

During the armistice negotiations, the two sides had different

였고 미국 내에서도 평화적인 휴전을 요구하는 여론이 고조되었습니다.*

6.25한국전쟁이 발발하고 약 1년이 지난 1951년 6월 23일 소련외상 제이고프 마리크(Jacob Malik)는 '평화의 대가' 라는 UN의 방송프로그램을 통해 공산측은 한국전쟁의 종식을 협상할 용의가 있음을 표명하였고, 1951년 6월 30일 유엔군 총사령관 마츄 비 리지웨이(Mathew B. Ridgeway) 대장은 'Jutlandia' 라는 라디오 프로그램을 통하여 휴전협상을 제의하였습니다.**

그 후 공산군측은 1951년 7월 1일 유엔군 총사령관의 제의를 수락하면서 7월 10일 정전협상을 개성에서 개최하고자 제안하였으며 이에 따라 7월 6일 쌍방이 연락장교 회의를 갖고 7월 10일부터 정전협상을 위한 본회의를 개최하게 되었습니다.

유엔군측과 공산군측은 6.25한국전쟁을 평화적으로 해결하기 위하여 1951년 7월 10일 첫 정전회담을 시작하였고, 1953년 7월 27일까지 약 2년간 한쪽에서는 치열한 전투를 그리고 다른 한쪽에서는 정전협상을 병행하였습니다.

유엔군측과 공산군측은 서로 적대행위 중지를 위하여 비무장지대를 설치하고 이를 위한 군사분계선 합의, 전투중지 및 정전협정 이행을 위한 구체적 조치, 전쟁포로에 관한 협상, 외국군대의 철수와 한반도 문제의 평화적 해결을 위한 정치회의 소집 등을 의제로 하는 유엔군측과 공산군측의 군사령관간에 이루어진 군사적 성격의 협상을 하였습니다.*** 그러나 오늘날까지도 정치회담을 통한 한국전쟁의 평화적 해결을 마무리하지는 못했습니다.

정전협정을 위한 협상은 "군사분계선을 38선으로 하

*이상철, 『한반도 정전체제』(서울 : 한국국방연구원, 2012), 11쪽

**이문항, 『JSA-판문점(1953~1994)』(서울 : 소하, 2001), 269쪽

UN General Assembly Adopting Resolutions

유엔총회 의결

***손규석 · 조성훈 · 김상원, 『6.25 전쟁과 유엔군』(서울 : 국방부 군사편찬연구소, 2015), 462쪽

opinions virtually on all issues and it took them 765 meetings over two years to sign an agreement. As for the MDL, the Communist side proposed the line be drawn along the 38th parallel whereas the UNC side insisted on the current points of contact being the demarcation line. On the POW issue, the UNC side wanted to give each POW the freedom to choose a destination while the Communist side demanded unconditional, blanket repatriations. Also, UNC argued for allowing media access to armistice negotiations, but the Communist side opposed to it citing confidential military information being discussed at the talks as a reason. In the end, the Korean Armistice Agreement was signed by the Commander of UNC, the Supreme Commander of KPA, and the Commander of CPV on July 27, 1953 (JSA Panmunjom, p. 289; outline and agenda of the truce negotiation).

As such, the Armistice Agreement is a truce agreement, it is military in nature and designed to cease all hostilities until such time that a peaceful resolution to the Korean War is finally achieved.

Signing the Armistice Agreement

자"는 공산군측과 "현재의 접촉선으로 하자"는 유엔군측 의견, 그리고 포로송환은 "포로의 자유의지에 따라 국가를 결정하자"는 유엔군측과 "무조건 송환을 주장"하는 공산군측, "언론보도인의 출입을 허용하자"는 유엔군측과 "휴전협상은 군사기밀을 취급하기에 기자들의 출입을 허가할 수 없다"는 공산군측 등 의제 하나하나마다 양측의 의견 차이가 매우 컸으며 이러한 상황에서 합의를 이끌어 내기 위하여 2년에 걸쳐 765회의 수많은 회의를 하였습니다. 그 결과 당시 유엔군 사령관과 조선인민군 최고사령관 및 중국인민지원군의 사령관은 1953년 7월 27일 마침내 정전협정을 체결하였습니다.*

이처럼 정전협정은 6.25전쟁의 평화적 해결이 최종적으로 달성될 때까지 적대행위를 완전히 정지하자는 군사적 성격의 협정, 즉 휴전협정을 한 것입니다.

• 접촉선(Line of Contact) : 남과 북한의 군사들이 대치하고 있는 접촉점.

정전협정 서명

* 이문항, 『JSA-판문점(1953~1994)』(서울 : 소하, 2001), 289쪽; 휴전협상의 의제와 협상개요

〈Establishing MDL and DMZ〉

The core spirit of the Armistice Agreement was to prevent the opposing sides from committing hostile acts and reinforcing military forces on the Korean Peninsula. The first step was to set up a MDL to separate the opposing forces, and establish a DMZ up to a certain distance from the demarcation line to create a buffer zone, and thereby prevent a resumption of hostilities.

The in-depth discussion on establishing the MDL and the DMZ began at the 11th plenary session on July 27, 1951, and ended at the 28th plenary session on November 27, 1951.

With regard to the MDL on the ground, the Communist side argued, "Given that the war broke out along the 38th parallel and given that both sides want to return to the state before the war, the 38th parallel should be the MDL and the areas up to 10km from the line on each side should be a demilitarized zone." On the contrary, UNC insisted on a 20 mile-wide demilitarized zone and argued, "The 38th parallel was drawn for political and historical reasons, and as such, is not suitable as the MDL. The MDL should not be an abstract line, but be a line reflective of the current, physical military situation" (JSA Panmunjom, p. 275).

Admiral Joy, who led the UNC delegation at armistice negotiations, said that since the start of the war, the opposing forces had crossed the 38th parallel back and forth at least four times. He also noted that not only the ground, but also the maritime and air battle zones should be taken into account when establishing the MDL, reminding the Communist side that the UN forces had a firm superiority in the air and at sea. His point was that when drawing the MDL on the ground,

〈군사분계선과 비무장지대 설치〉

정전협정의 핵심은 유엔군측과 공산군측 쌍방간에 적대행위를 금지하고 한반도에 새로이 군사력을 증강하지 못하도록 하는 것이었습니다. 먼저 쌍방의 군사력을 격리시키기 위하여 군사분계선을 설정하고 군사분계선으로부터 일정한 폭을 비무장지대로 설정하여 이를 완충지대로 함으로써 적대행위의 재발을 방지토록 하는 것이었습니다.

군사분계선과 비무장지대
MDL and DMZ

군사분계선과 비무장지대의 설정문제는 1951년 7월 제11차 본회의에서부터 1951년 11월 제28차 본회의까지 협의를 집중적으로 하였습니다.

지상 군사분계선과 관련하여 공산군측은 "38도선에서 전쟁이 일어났으므로 원점으로 돌아가기 위해서는 38도선을 군사분계선으로 설정하고, 그 38도선을 중심으로 쌍방 10km의 비무장지대를 설치할 것"을 주장하였습니다. 하지만 유엔군측은 "정치 · 역사적 이유로 설정된 38도선은 군사분계선으로 적합하지 않는 바, 군사분계선은 추상적인 선이 되어서는 안 되며 현존하는 실질적인 군사 상황에 준하는 선이 되어야 한다"고 하면서, 20마일의 비무장지대를 제시하였습니다.*

유엔군 수석대표 조이 제독은 전쟁 이후 지금까지 쌍방이 적어도 네 번이나 38선을 넘어갔다왔다 했으며, 군사분계선의 설정에는 지상뿐만 아니라 해상과 공중전투 구역도 고려해야 한다면서 당시 해 · 공군의 패권을 유엔군이 갖고 있음을 상기시켰습니다. 즉 지상에서 분계

*이문항, 『JSA-판문점(1953~1994)』(서울 : 소하, 2001), 275쪽

the Communist side should make concessions to compensate for UNC's air and naval superiority. In fact, UNC used its superiority as a leverage to secure the Kaesong area, but in vain. The Communists called the UNC's claims nonsense and kept insisting on the 38th parallel constituting the MDL. They also insisted that the UN forces be withdrawn from those islands in the rear areas of North Korea (JSA Panmunjom, p. 80).

Finally, after numerous negotiations and a month-long closed-door sessions, the Communist side signaled that it could agree to the UNC's terms if it no longer used its air and naval superiority as a bargaining chip to demand additional concessions. The UNC side also offered that it could agree to some of the claims by the North such as the width of the DMZ. As a result, a comprehensive agreement was reached at the 28th plenary session; the Communist side conceded to have the current line of contact be the MDL on the ground and allow the Neutral Nations Supervisory Commission (NNSC) to inspect the North Korean areas, and in return, the UNC side agreed to withdraw its forces from those islands that had been under the North Korean control prior to the war and to have a 4km-wide demilitarized zone. This is how the MDL and the DMZ of today were created (The Korean Peninsula Armistice Regime, p. 140).

As prescribed in Paragraph 4, Article I of the Korean Armistice Agreement, markers were erected to delineate the MDL, which was completed in September 1954. A total of 1,292 markers were erected along the entire 155 mile-long MDL with the first marker, MDL Marker No. 0001, at Jangdan Peninsula at the estuary of the Imjin River and the last one, MDL Marker No. 1292, on the East coast. Of those, 696 markers were assigned to UNC while the remaining 596 were assigned to the Communist side for maintenance (The Korean

선을 그을 때는 공산군측이 그만큼 유엔군측에 대가를 치러야 할 것이라고 언급하였습니다. 유엔군측은 우세한 해 · 공군력을 바탕으로 개성지역을 확보하려고 했었습니다. 그러나 공산군측은 터무니없는 소리라고 하면서 38선을 계속 주장하였고, 북한 후방지역의 이북도서에서 유엔군이 철수할 것을 요구하였습니다.*

*이문항, 『JSA-판문점(1953~1994)』(서울 : 소하, 2001), 80쪽

결국 수많은 협의와 1개월여 간의 비공개 회의 끝에 공산군측은 만일 유엔군측이 공군과 해군의 우세한 전력에 따른 추가요구를 하지 않는다면 유엔군측의 주장에 동의할 수 있음을 암시하였고 유엔군측은 비무장지대의 넓이 등 북측의 주장에 동의할 수 있음을 제기하였습니다. 이에 제28차 본회의에서 포괄적으로 합의가 이루어졌는 바 공산군측은 현 전선의 접촉선을 지상군사분계선으로 설정하고 중립국감독위원회의 북한지역 사찰을 허용하며 그 대가로 유엔군측은 전쟁 이전 북한이 관할하던 이북 도서들에서 철수하고 비무장지대는 넓이를 4km로 하자는 데 양측은 합의하였고 바로 이것이 오늘날의 군사분계선과 비무장지대가 되었습니다.**

- 접촉선(Line of contact) : 남과 북한의 군사들이 대치하고 있는 접촉점.
- Neutral Nations Supervisory Commission(NNSC) : 중립국감독위원회

**이문항, 『JSA-판문점(1953~1994)』(서울 : 소하, 2001), 275쪽; 이상철, 『한반도 정전체제』(서울 : 한국국방연구원, 2012), 140쪽

정전협정 1조 4항에 의거 1954년 9월까지 군사분계선에 표식작업을 하였으며, 임진강 하구 장단반도에 세워진 표식물 제0001호에서 시작하여 동해안의 마지막 표식물 제1,292호까지 155마일(약 250km)에 걸쳐 표식을 하였습니다. 그중 696개는 유엔군측이, 나머지 596개는 공산군측이 관리책임을 맡기로 하였습니다.***

***이상철, 『한반도 정전체제』(서울 : 한국국방연구원, 2012), 137~138쪽

Peninsula Armistice Regime, pp. 137~138).

The marking of the MDL is important because it serves not only as the line separating the South and the North, but also the baseline for all military operations of both sides.

The MDL markers were painted yellow with black lettering in Korean and English on the side facing south and in Korean and Chinese on the side facing north and a serial number at the bottom.

The distance between each marker should be no more than 500 meters in a straight line, or 300 meters in a curved line to ensure the adjacent markers on its left and right could be seen from any marker. Contrary to the common belief that the MDL is a barbed wire fence line, it is an invisible line marked with steel markers. Many of the MDL markers have been lost due to wear and tear, floods and poor maintenance, becoming a source of friction between the two Koreas that resulted in many casualties. For instance, on March 12, 1969, a North Korean guard post shot at four U.S. soldiers who were replacing MDL Marker No. 0022, killing one and wounding the other three. What was even more tragic was that the U.S. helicopter that was evacuating them crashed, killing eight more people(JSA Panmunjom, p. 84).

With another firing incident by North Korea in 1973 during the maintenance of MDL Marker No. 654, the Commander of UNC ordered the suspension of the marker maintenance on March 7, 1973 after providing the North with a prior notification(Military Armistice Commission Yearbook Vol. 8, p. 87).

Although the MDL was established on the ground, UNC and the Communist side were unable to reach an agreement on a demarcation line at sea.

이 군사분계선의 표식작업이 중요한 이유는 남북을 가르는 선이 될 뿐만 아니라 남북군사작전의 기준이 되기 때문입니다.

표식판의 바탕은 황색이고 글자는 흑색이며 남쪽을 향해서는 국문과 영문으로 북쪽을 향해서는 국문과 중국어로 '군사분계선' 이라고 쓰고 그 밑에는 표식판 번호가 표기되었습니다.

그리고 표식판의 간격은 좌우로 보았을 때 식별이 가능해야 하며 직선일 경우는 500미터를 넘지 않고 곡선일 경우는 300미터를 넘지 않는 거리를 유지토록 하였습니다. 군사분계선을 넘었다고 하면 마치 철책선을 넘은 것으로 이해할 수 있으나 사실 군사분계선은 표식판으로 표식되어 있습니다. 많은 표식판들이 낡아서 손실되고, 홍수로 떠내려가고 보수도 잘 되지 않아 이로 인한 남북한의 마찰로 많은 사상자가 발생하기도 하였습니다. 그 중 한 예를 보면 1969년 3월 12일 미군 4명이 표식판 0022호를 새로 갈아 끼우고 돌아서자 북한측 초소에서 총격을 가해서 미군 1명이 사망하고 3명이 부상을 입었고 이를 후송하려던 미군 헬기가 추락해서 8명이 더 사망한 안타까운 사건도 있었습니다.*

군사분계선 표식판 보수작업은 1973년 표식판 654번 보수작업 중 북측의 총격사건이 있었고 1973년 3월 7일 북측에 사전 통보 후 유엔군 사령관 명에 의거 표식판 보수작업은 중단되었습니다.**

지상에서의 군사분계선은 설정되었으나 해상에서의 군사분계선은 공산군측과 유엔군측의 의견이 달라 합의되지 못했습니다.

군사분계선 표식물
MDL Marker

군사분계선 표식물
MDL Marker

* 이문항, 『JSA-판문점(1953~1994)』(서울 : 소하, 2001), 84쪽

** 합참정보본부, 『군사정전위원회 편람』(제8집), 87쪽

At that time, the two sides maintained an unbridgeable gap on maritime boundaries as the UN forces, which had full control over the air and sea of the Korean Peninsula, insisted on the three nautical mile rule, which was the international norm at the time for territorial waters, for the maritime MDL, but the Communist side insisted on twelve nautical miles out of fear for maritime blockades by the UN forces(The Korean Peninsula Armistice System, p. 142).

The UNC side argued that the North had nothing to worry about as maritime blockades were specifically prohibited in Paragraph 15 of the Armistice Agreement, but the Communist side demanded the clauses pertaining to territorial waters that were being negotiated be removed, which the UNC side eventually accepted. As such, no maritime demarcation lines were agreed by the two sides(The Korean War and UN Forces, p. 468).

In an effort to prevent armed clashes between the South and the North, however, the Commander of UNC, Mark W. Clark, who had the responsibility and authority to implement and comply with the Armistice Agreement, established the Northern Limit Line (NLL) on August 30, 1953 to set the boundary for patrol operations for UNC's ships and aircraft (ROK Ministry of National Defense (MND)'s Position on NLL, p. 7).

The Northern Boundary Line (NBL) was drawn in the East Sea, which was an extension of the ground MDL into the sea. As for the NLL in the West Sea, the international norm of the three nautical mile rule for territorial waters was applied and the NLL was drawn in between the five Northwest Islands and the North Korean territories and twelve grid coordinates were connected from the Han River Estuary (HRE) to northwest of Baengnyeong-Island. It was because North Korea's naval forces were practically nonexistent at the time,

당시 한반도 전역에 대한 제공권과 제해권을 장악하고 있던 유엔군측은 해상에서의 군사분계선을 당시 국제적으로 통용되는 3해리(NM) 영해를 주장하였고, 공산군측은 유엔군측의 해안봉쇄를 우려하여 12해리(NM) 영해를 주장하여 쌍방의 의견 차이가 컸습니다.*

*이상철, 『한반도 정전체제』(서울 : 한국국방연구원, 2012), 142쪽

유엔군측은 정전협정 제15항에 해안봉쇄를 하지 못한다는 조항이 포함되어 있어 문제가 없다고 주장하였으나 공산군측은 당시 협의되고 있었던 영해 설정 관련조항 삭제를 요구하였고 이를 유엔군측이 수용함으로써 해안에서의 경계선 설정은 합의에 실패하였습니다.**

**손규석 · 조성훈 · 김상원, 『6.25 전쟁과 유엔군』(서울 : 국방부 군사편찬연구소, 2015), 468쪽

그러나 1953년 8월 30일 정전협정을 이행하고 준수해야 할 책임과 권한을 가진 유엔군사령관 마크 더블유 클라크(Mark W. Clark)는 남북간의 무력충돌을 방지하기 위하여 유엔군측 함정과 항공기의 해상 초계활동 범위에 대한 북방한계선(NLL; Northern Limit Line)을 설정하였습니다.***

***국방부, 『북방한계선에 관한 우리의 입장』(서울 : 신오성기획인쇄사, 2007), 7쪽

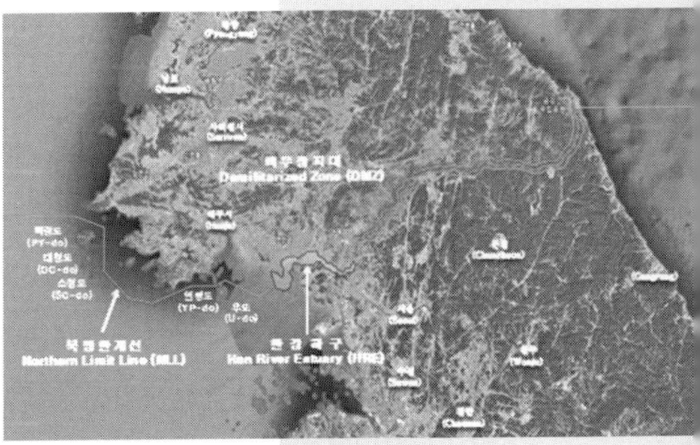

북방한계선
The Northern Limit Line

동해는 지상 군사분계선의 연장선을 기준으로 북방경계선(NBL: Northern Boundary Line)을 설정하였고 서해의 북방한계선(NLL: Northern Limited Line)은 당시 국제적으로 통용되던 폭 3해리 영해에 입각하여 서북 5개 도서들과 북한지역과의 대략적인 중간선을 기준으로 한강 하구로부터 백령도 서북방까지 12개 좌

and the UN forces controlled the entire waters surrounding the Korean Peninsula. In other words, armed clashes between the two sides at sea could be prevented simply by controlling activities of the UN forces. The names of the lines in the East and West Seas later became unified as the NLL on July 1, 1996 (The Korean War and UN Forces, p. 469; ROK MND's Position on NLL, p. 8).

〈Concrete Arrangements for Armistice: Military Armistice Commission (MAC) and Neutral Nations Supervisory Commission (NNSC)〉

In the simplest terms, the core obligation of the Armistice Agreement is to prevent both sides from engaging in hostilities and introducing additional military forces onto the Korean Peninsula. To this end, the DMZ was established based on the MDL to separate the opposing forces and the two sides are obligated to respect areas under the control of the other side. To uphold the spirit of the Armistice Agreement, procedures are in place to ensure compliance and to control access to the DMZ. To fulfill these core obligations, established were the NNSC, and the MAC which is jointly represented by the UNC side and the Communist side.

The MAC oversees the DMZ and the HRE to prevent hostilities, and the NNSC monitors the movement of personnel and equipment at five designated ports and airports of entry outside the DMZ and HRE in North and South Korea each to ensure no additional military forces are introduced to the Korean Peninsula (Paragraphs 27 and 42, Armistice Agreement).

표를 연결하여 설정하였습니다. 당시 북한군의 해군세력이 거의 전무하였고 유엔군이 한반도의 전 해역을 장악하고 있었던 상황에서 유엔군측의 활동만 적절히 통제하면 쌍방간의 무력충돌을 예방할 수 있었기 때문입니다. 그 이후 1996년 7월 1일 명칭을 동 · 서해 모두 북방한계선(NLL)으로 통일하여 오늘날에 이르고 있습니다.*

*손규석 · 조성훈 · 김상원, 『6.25 전쟁과 유엔군』(서울 : 국방부 군사편찬연구소, 2015), 469쪽; 국방부, 『북방한계선에 관한 우리의 입장』(서울 : 신오성기획인쇄사, 2007), 8쪽

〈정전의 구체적 조치 : 군사정전위원회, 중립국감독위원회〉

정전협정의 핵심 의무사항은 쌍방이 적대행위를 하지 못하도록 하고 한반도에 새로이 군사력을 증강하지 못하도록 하는 것이었습니다. 따라서 양측 군대를 분리하기 위하여 군사분계선을 기준으로 비무장지대를 설치하고 상대측 지역을 서로가 존중하는 것입니다. 또한 정전협정을 준수하기 위한 절차를 수립하고 비무장지대의 출입을 제한하여 정전협정의 정신을 준수하는 것입니다. 이러한 핵심 의무사항을 이행하기 위하여 중립국감독위원회와 유엔군측과 공산군측 대표로 공동 구성되는 군사정전위원회가 설립되었습니다.

군사정전위원회는 비무장지대와 한강하구지역을 감독하면서 적대행위를 하지 못하도록 감독하고 중립국감독위원회는 비무장지대와 한강하구지역 이외 남북한 지역에 각각 지정된 5개 항구와 공항에서 인원 및 장비의 이동을 감시하며 한반도에 새로이 군사력을 증강하지 못하도록 하였습니다.**

**정전협정 27, 42항

Military Armistice Commission

The MAC is a joint body between UNC and the Communist side with a total of ten senior officers from both sides without a chairperson. The Commander of UNC appoints five officers and the other five are appointed jointly by the Supreme Commander of KPA and the Commander of CPV (Paragraph 20, Armistice Agreement).

In general, the MAC oversees the implementation of the Armistice Agreement and settles through negotiations any Armistice violations. Its mission includes revising the Armistice Agreement, establishing procedures and regulations for the MAC meetings, providing general supervision and direction to the Committee for Repatriation of Prisoners of War and the Committee for Assisting the Return of Displaced Civilians, operating Joint Observation Teams (JOTs), and maintaining communication channels between the two opposing commanders (Paragraphs 24 and 25, Armistice Agreement).

To observe activities in the DMZ, the MAC operated ten JOTs and did not allow anyone to cross the MDL unless specifically approved. Also, the MAC routinely inspected the DMZ, took corrective actions as needed, and deployed a JOT for an on-site investigation if suspicion arose(Paragraphs 26 and 27, Armistice Agreement).

After the signing of the Armistice Agreement in 1953, the MAC established and operated ten JOTs to supervise the withdrawal of military forces and the removal of all Explosives, minefields and wire entanglements from the DMZ, but the number of JOTs was reduced to six in 1954 and then to five in 1955. Each JOT was assigned to a certain sector of the DMZ: JOT 1 to the HRE area, JOT 2 to the area in between MDL Markers 0001 and 0343, JOT 3 to the area in between MDL

군사정전위원회

군사정전위원회는 위원장이 없는 유엔군측과 공산군측 대표 공동조직체로서 총 10명의 고급장교로 구성되었으며 그 중 5명은 국제연합군 총사령관이 임명하며 나머지 5명은 조선인민군 최고사령관과 중국인민군지원군 사령관이 공동으로 임명토록 하였습니다.*

군사정전위원회의 전반적 임무는 정전협정의 이행을 감독하며 정전협정의 어떠한 위반사건이 발생하면 협의하여 처리합니다. 그리고 정전협정의 수정에 관한 사항, 군사정전위원회 회의주관과 필요한 절차 및 규정의 채택, 전쟁포로송환위원회와 실향사민귀향협조위원회의 사업의 감독과 지도, 공동감시소조의 운영, 쌍방 사령관들 간의 대화 통로 유지 등의 임무를 수행토록 하였습니다.**

특히 비무장지대 감시 임무를 수행하기 위하여 군사정전위원회는 10개의 공동감시소조를 두어 그 협조를 받으며 특정한 허가 없이는 누구도 군사분계선을 통과하지 못하도록 하고 주기적으로 비무장지대를 점검하여 시정조치를 하고, 의심사항 발생시 공동감시소조를 파견해 현지조사를 실시토록 하였습니다.***

정전협정 이후 1953년에 군사정전위원회는 10개의 공동감시소조를 편성 운영하여 비무장지대로부터의 부대철수와 모든 폭발물과 지뢰 그리고 철조망의 제거 여부를 감독하는 일을 수행하였으나, 1954년에 6개 소조, 1955년에는 5개 소조로 축소되었으며 당시 담당구역은 제1조는 한강하구, 제2조는 표식물 0001호부터 0343호, 제3조는 제2조 동단부터 표식물 0710까지, 제4조는 제3

*정전협정 20항

**정전협정 24, 25항

• 공동감시소조 : 4~6명의 영관장교로 구성, 쌍방이 50% 임명. 비무장지대와 한강하구 지역의 정전협정 이행을 감독.

***정전협정 26, 27항

Markers 0343 and 0710, JOT 4 to the area in between MDL Markers 0710 and 0982, and JOT 5 to the area in between MDL Markers 0982 and 1292. There had been a total 279 JOT meetings held by the end of 1965, but JOTs began to slowly weaken with North Korea's maneuver to undermine the MAC. North Korea began boycotting the MAC in March 1991 when it was notified of the appointment of a ROK major general, Hwang Won-tak, as the Senior Member of UNCMAC. Eventually it withdrew its Communist delegation from the MAC in 1994, and with this event, the MAC operation in the North practically came to a halt.

Although the JOT was requested on over 80 occasions, it only stood up a few times. When investigating straightforward cases such as an accidental crossing of the MDL, the JOT was able to conclude its investigation with ease with the violating side owning up to the mistake. However, when it came to more complex cases such as North Korea's fortifications in the DMZ, dispatching armed infiltrators, and

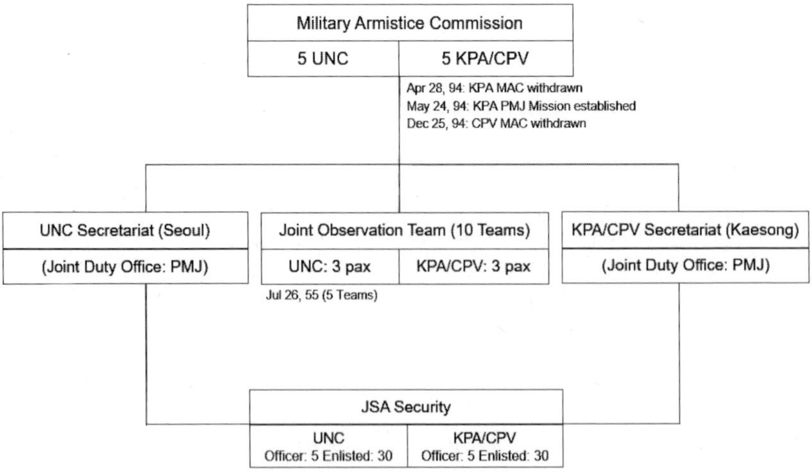

조의 동단부터 표식물 0982까지, 제5조는 제4조의 동단부터 표식물 1292호까지로 담당하였습니다. 이후 1965년 말까지 총 279회의 공동감시소조 회의가 있었으나 북측의 군사정전위원회 무력화로 조금씩 부실화되기 시작하였고, 1991년 3월 황원탁 한국군 소장을 군사정전위원회 수석대표로 임명하고 이를 북측에 통보하자 북측은 군사정전위원회를 보이콧하기 시작하였고, 1994년 공산군측이 군사정전위원회로부터 철수함으로써 사실상 북측의 기능이 거의 마비되었습니다.

공동감시소조의 구성은 80여 차례 요구가 있었으나 실제로 구성된 것은 몇 차례 되지 않으며 공동조사결과 사례를 보면 군사분계선을 잘 알아보지 못한 단순 월경사건 등 단순한 사건은 쉽게 조사를 끝내고 실수를 시인했으나, 비무장지대 안에 구축한 북한의 요새 진지, 무장간첩 침투사건, 쌍방의 총격교전사건 등을 공동조사한 결

유엔-북한군 공동감시소조

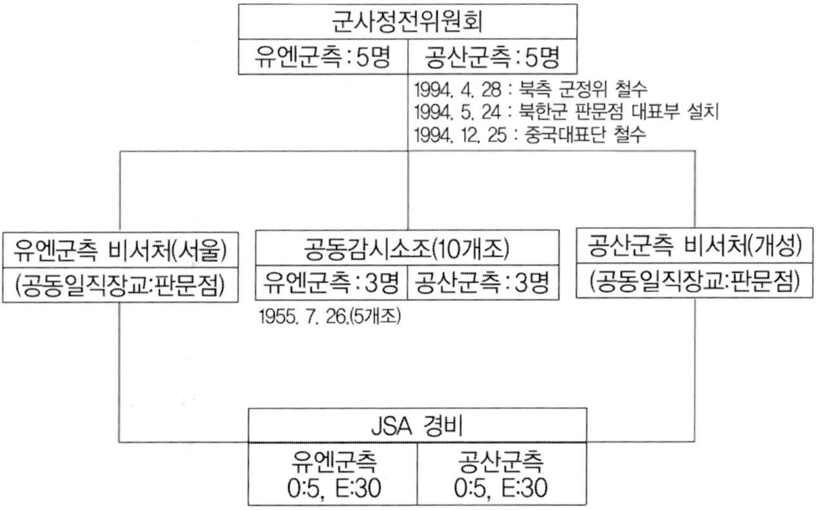

exchanging fire, the two sides stuck to their stories, failed to find a middle ground, and almost always, submitted exact opposite reports to the MAC. (Military Armistice Commission Yearbook Vol. 8, p. 98).

Due to the KPA's reluctance to participate in the JOT, the Commander of UNC directed the Special Investigation Team (SIT) to conduct unilateral investigations in 1968, and the SIT has been in operation ever since (Military Armistice Commission Yearbook Vol. 8, p. 99).

The MAC met at the Joint Security Area in Panmunjom in the DMZ at the request of either party and they met in the forms of the MAC Meeting, Secretary Meeting, Staff Officer Meeting and Joint Duty Officer Meeting*. The MAC meeting oversaw the implementation of the Armistice Agreement and discussed/resolved major Armistice violations such as violating the DMZ, infiltrating into the air and maritime spaces of the other side, augmenting military forces, sending infiltrators, building tunnels, shooting down aircraft, and sinking fishing vessels. There were 459 MAC meetings from July 28, 1953 and to February 1991. Whenever an Armistice violation occurred, the MAC meeting was held and lasted for several hours, but usually to no avail as the two sides stuck to their positions and refuted claims by the other side. Nonetheless, the MAC meeting offered a mechanism for dialogue that provided opportunities to understand North Korea's intentions on pending issues, deterring all-out war between the North and the South, They also helped to define issues friction for each side whether AA violations or not, and contributing to stability on the Korean Peninsula (The Korean War and UN Forces, p. 472).

The Secretaries of the MAC held 510 meetings between July 27, 1953 and June 6, 1994, to discuss and handle minor Armistice violations and matters concerning repatriation. Currently, the Secretary of the UNC

과에 대해서는 쌍방이 자기 주장만 하였고 동의하지 못해 거의 언제나 정반대의 공동감시소조 보고서를 군사정전위원회에 제출하였습니다.*

공동감시소조 참여에 대한 북한군측의 미온적 태도로 인해 유엔군사령관은 1968년부터 유엔군사령부 특별조사반(SIT)에 의한 일방적인 조사를 지시하여 현재까지 운용되고 있습니다.**

군사정전위원회 회의는 어느 일방의 요청에 의해 비무장지대내에 있는 판문점 공동경비구역에서 실시되며 본회의, 비서장회의, 참모장교회의, 공동일직장교회의 등이 있었습니다. 본회의는 정전협정의 이행을 감독하고 비무장지대 위반, 공중 및 해상침범, 무력증강, 간첩침투, 땅굴사건, 항공기 격추사건, 어선 격침사건 등 주요 정전협정 위반사항에 대해 협의 · 처리해 왔으며, 1953년 7월 28일 제1차 회의를 시작한 이후 1991년 2월까지 총 459차례 개최되었습니다. 회의는 정전협정 위반사항이 발생시마다 몇 시간씩 진행되었으나 쌍방이 자기 주장과 반박으로 별 성과를 내지 못했습니다. 그럼에도 불구하고 군사정전위원회 회의는 의사소통기구로서 현안에 대한 북한측 의도를 파악할 수 있었으며, 또한 정전협정 위반 여부의 갈등을 야기시키는 쟁점 사안을 규정하는 데 도움이 되었고, 남북한간의 전면적인 전쟁을 억제하여 한반도의 안정에 기여하여 왔습니다.***

군사정전위원회 비서장회의는 정전협정의 경미한 위반행위에 대한 처리, 유해송환에 관한 업무 등을 협의 · 처리하고 1953년 7월 27일부터 1994년 6월 6일까지 총

* 합참정보본부, 『군사정전위원회 편람』(제8집), 98쪽

• SIT : Special Investigation Team

** 합참정보본부, 『군사정전위원회 편람』(제8집), 99쪽

• 본회의 : 소장급을 수석위원으로 하는 쌍방 5명씩의 위원으로 구성.

*** 손규석 · 조성훈 · 김상원, 『6.25 전쟁과 유엔군』(서울 : 국방부 군사편찬연구소, 2015), 472쪽

• 비서장회의 : 쌍방 비서장(대령급)으로 구성되며 부비서장 및 관계 요원 10명이 배석.

side of the Military Armistice Commission (UNCMAC) maintains contact with the representative of the KPA Panmunjom Mission (The Korean Peninsula Armistice Regime, p. 38).

UNCMAC is composed of members from the U.S., ROK, United Kingdom (U.K.) and Sending States, which take turns to fill the Rotating Member position. The position of the Senior Member used to be held by a U.S. general officer after the signing of the Armistice Agreement. As previously mentioned was transferred to Rok MG. since of March 25, 1991.

At the time, the member of the KPA/CPV MAC stated that it was unacceptable to have a South Korean general officer as the Senior Member as South Korea was unofficially opposed to signing of the Armistice Agreement and didn't even exercise operational control over its forces. He threatened that the MAC and the NNSC would cease to function if a South Korean general officer became the Senior Member (JSA Panmunjom, p. 225).

The Communist side of the MAC consisted of KPA and Chinese delegations. In March 1991, however, with the appointment of a ROK general officer as the Senior Member of UNCMAC, the Communist side refused to attend MAC meetings and continued to undermine the MAC with an ultimate goal to dismantle the Armistice regime. Furthermore, they withdrew the KPA delegation from the MAC in April 1994 and the Chinese delegation in December of 1994.

Despite this, in May 1994, North Korea established the KPA Panmunjom Mission and called for talks for a peace treaty between the U.S. and North Korea.

510회 개최되었습니다. 현재는 유엔군사령부 군사정전위원회 비서장과 조선인민군 판문점 대표부 연락관과 접촉을 유지하고 있습니다.*

군사정전위원회 유엔군측 위원은 미국, 한국, 영국, 그리고 참전국 연락장교단 대표들을 순서에 의거하여 순환대표로 임명하고 있으며, 수석대표는 정전협정 체결 이후 미군 장성이 임명되어 오다가 앞에서 언급한 바와 같이 1991년 3월 25일 이후 한국군 장성이 임명되어 오고 있습니다.

당시 북한측 군사정전위원은 비공식적으로 남측은 정전협정체결을 반대하였고, 작전권한도 없는데 한국군 장성을 수석대표로 임명하는 것은 받아들일 수 없고 그러한 일이 감행된다면 군사정전위원회나 중립국감독위원회는 마비될 것이라고 언급하였습니다.**

* 이상철, 『한반도 정전체제』(서울 : 한국국방연구원, 2012), 38쪽

** 이문항, 『JSA-판문점(1953~1994)』(서울 : 소하, 2001), 225쪽

군사정전위원회 공산군측은 북한군, 중공군으로 구성되어 있었으나, 1991년 3월 유엔사측 수석대표로 한국군 장성이 임명된 후 공산군측은 본회담 참석을 거부하였고, 정전체제를 와해시키려는 의도로 공산군측 군사정전위원회를 무력화시켜 왔습니다. 북한은 1994년 4월 군사정전위원회 북한군 대표들을 철수시켰고, 12월에는 중공군 대표들까지 철수시켰습니다.

반면에 북측은 1994년 5월 조선인민군 판문점 대표부를 설치하면서 미·북 평화협정 체결을 위한 회담을 개최할 것을 주장하였습니다.

〈North Korea's Rationale for Refusing to Acknowledge South Korean Officer as Senior Member and to Participate in MAC Meetings〉*

1. The South Korean military is neither a party to the Armistice Agreement nor a member of the UN forces, and South Korea was opposed to the signing of the Armistice Agreement.
2. The South Korean military does not represent the entire armed forces in South Korea.
3. A South Korean general officer is not eligible to become the Senior Member of the UNC side of the Military Armistice Commission, which implements and monitors the Armistice Agreement.
4. Due to the illicit appointment of the Senior Member, the Military Armistice Commission cannot conduct its normal activities.

With such actions by North Korea, the MAC was disabled, paralyzing the dialogue channel for managing crises on the Korean Peninsula. What this meant was that there wouldn't even be an opportunity to seek a peaceful resolution to prevent escalation in the event of a major hostile act such as physical provocations and accidental armed clashes between the South and the North in the DMZ. Meanwhile, a U.S. military helicopter was shot down north of the MDL on December 17, 1994 and the necessity of having a functioning MAC was highlighted in the process of negotiating the repatriation of the U.S. pilots. With such recognition, the UNC and the ROK MND worked closely to develop the procedures for the UNC-KAP General Officer-level Talks (draft) in 1995 (The Korean Peninsula

〈북한의 군사정전위원회 한국군 수석대표 불인정 및 본회의 참석 거부 이유〉*

*Announced by KCNA on March 28 as North Korea refused to accept the Senior Member accreditation letter on March 25, 1991

*1991년 3월 25일 수석대표 신임장 접수 거부시, 3월 28일 조선중앙통신

1. 한국군은 정전협정의 당사자도 아니고 유엔군의 일원도 아니며, 남한은 정전협정 체결을 반대했음.
2. 한국군은 남한내의 무력 전체를 대표하지 않음.
3. 한국군 장성은 정전협정을 이행 · 감시하는 군사정전위원회 유엔군측 수석대표가 될 수 없음.
4. 수석대표의 비합법적 임명으로 군사정전위원회의 정상적인 활동을 할 수 없음.

이처럼 북한에 의해 군사정전위원회가 제 기능을 발휘할 수 없게 되었는데 이는 한반도 위기관리를 위한 협의통로가 마비되었다는 것을 의미했습니다. 예를 들어 비무장지대 내 총격도발, 남북간 우발적 무력충돌 등 중대한 적대행위가 발생할 경우 평화적 해결을 통해 위기가 확대되는 것을 방지할 수 있는 기회조차 상실하게 되었던 것을 의미합니다. 그러던 중 1994년 12월 17일 미군헬기가 군사분계선을 월선하여 피격 · 추락하는 사고가 발생하고 미군 조종사 송환문제를 해결하는 과정에서 군사정전위원회가 정상적으로 가동되어야 할 필요성을 인식하고 1995년부터 유엔사와 한국 국방부는 긴밀한 협의를 통해 유엔사-조선인민군간 장성급회담을 위한 절

Armistice Regime, p. 39).

After several Secretary and staff officer-level talks, the two sides agreed on the "Procedures for General Officer-level Talks between UNC and KPA" within the framework of the Armistice Agreement on June 8, 1998, which has replaced the functions of the MAC.

〈Procedures for UNC-KPA General Officer-level Dialogue on Armistice Issues〉

1. Title of Talks: Dialogue between General Officers of UNC and KPA concerning Armistice issues
2. Date of the Talks: The initial meeting shall be held on a date to be determined in conjunction with agreement on the stipulations for the dialogue and subsequent meetings shall be held at the request of either side.
3. Venue for the Talks: As Panmunjom on the MDL in the third building from the east end of Conference Row, Building T-2.
4. Agenda for the Talks: The Armistice Agreement as the Basis for Conflict Resolution and Prevention on the Korean Peninsula
   a. Resolving Armistice-related incidents involving military forces of the two sides.
   b. Resolving disagreements on interpretations of the provisions of the Armistice Agreement and the Subsequent Agreements
5. Participants
   a. Each side shall be allowed to designate a maximum of four

차(안) 마련을 추진하였습니다.*

1998년도에 군사정전위원회 비서장 및 참모장교간 수차례 협의를 거쳐 6월 8일 정전협정 틀내에서 "유엔사–조선인민군간 장성급회담 개최를 위한 절차"에 합의하였고 이는 군사정전위원회의 기능을 대신하게 되었습니다.**

* 이상철, 『한반도 정전체제』(서울 : 한국국방연구원, 2012), 39쪽

** 정전협정 서명 후에 이루어진 합의사항(U)

〈정전문제에 관한 유엔군과 조선인민군간의 장성급 대화를 위한 절차〉

1. 회담 명칭은 정전문제에 관한 유엔군과 조선인민군 장군들간의 대화.
2. 회담 일자는 첫 번째 회의는 회담을 위한 규정들에 대한 합의와 관련하여 결정되는 날짜에 개최하고, 그 후의 회담들은 어느 일방의 요청에 따라 개최한다.
3. 회담 장소는 판문점 군사분계선상에 있는 회담장 구역 동쪽 끝에서 3번째 건물 T-2임.
4. 회담 의제는 한반도에서 충돌을 방지하고 해결하기 위한 근거인 정전협정을 의제로 한다.
   가. 쌍방의 군대가 개입된 정전 관련 사건들의 해결
   나. 정전협정과 추후 합의사항들의 해석에 대한 이견 해소
5. 참석자
   가. 쌍방은 각자 최대 4명의 대표를 지명할 수 있

delegates. Each side shall have the right to select its delegates.

b. All delegates and one interpreter from each side shall be seated at the main conference table.

c. All delegates shall have an equal voice.

d. The senior officer of each side shall be head of delegation.

e. All delegates shall retain the right to attend all meetings between the two sides.

Each side shall be allowed to have additional assistants as participants in the General Officer-level meetings. These assistants will not sit at the main conference table and will not make statements during formal sessions.

6. Form of Talks:

a. The talks will be closed-door. No press will be allowed to view a meeting unless previously agreed upon by both sides.

b. The results of any meeting may be made public by either side without prior coordination with the other side.

Since the UNC and the KPA agreed on the procedures for the General Officer-level talks, there have been dozens of such talks between the two sides. The main agenda ranged from reducing tensions on the Korean Peninsula and defining accountabilities, to raising complaints against the North Korean Yugo-class submarine's infiltration into the ROK territorial waters in 1996, and to signing an agreement on opening some area of the DMZ to connect Gyeongui

도록 허용하고 각자의 대표들을 선정할 권리를 갖는다.

나. 쌍방의 대표자 전원과 양측에서 각각 통역 1명이 주 회의석상에 자리한다.

다. 쌍방의 대표자 전원은 동등한 발언권을 갖는다.

라. 쌍방의 선임장교가 각기 자기측 대표단의 선임자 역할을 한다.

마. 쌍방의 대표자 전원은 모든 회의에 참석할 권리를 갖는다.

쌍방은 추가적인 보조인원들을 장군급 대화에 참석시킬 수 있도록 허용한다. 이 보조인원들은 주 회의석상에 앉지 않으며 동 회의 중 발언을 하지 않는다.

6. 회담형식

가. 회담은 비공개로 한다. 쌍방이 사전에 합의하지 않는 한 보조인원들의 회의 참관은 허용하지 않는다.

나. 모든 회의결과는 상대방과의 사전 협의 없이 각자 공표할 수 있다.

유엔군과 조선인민군간 장성급 회담 절차 합의 이후 현재까지 십여 차례 장성급 회담을 하였습니다. 회담 주요내용은 한반도 긴장완화 및 책임문제, 1996년 북한의 유고급 잠수함 한국영해 침범에 대한 항의, 경의선 철도연결과 도로건설관련 비무장지대 일부 구역 개방에 대한 유엔사와 북한군간 합의서 서명, 동해선 철도연결 및 도로건설 관련 비무장지대 일부 구역 개방에 대한 유엔

railways and roads. Also, the two sides have continued to exchange messages and hold talks at the staff officer and colonel level.

NNSC : Neutral Nations Supervisory Commission

The mission of the NNSC was to prevent the North and the South from introducing additional military forces after the signing of the Armistice Agreement. To this end, the NNSC stationed Neutral Nation Inspection Teams (NNITs) at the ten designated areas in North and South Korea to monitor and inspect activities there and conducted special surveillance and inspections when violations occurred (Paragraph 42, Armistice Agreement). The number of personnel of the NNSC delegations peaked in 1953 at around 100 each for the Swiss and Swedish delegations and around 300 each for the Polish and Czechoslovakian delegations. As the NNSC's functions and activities decreased after 1956, the number at each NNSC delegation dropped sharply to about a dozen. Now there are only five officers each on the NNSC delegations on the UNC side(The Korean War and UN Forces, p. 477).

사와 북한군간 합의서 서명 등이었습니다. 또한 유엔군과 조선인민군간 메시지를 송수신하고, 참모장교급 회담과 대령급 회담을 지속적으로 해 왔습니다.

판문점 중립국감독위원회

중립국감독위원회

중립국감독위원회의 임무는 정전협정 후 남측과 북측이 새로이 군사력을 증강시키는 것을 감시하는 것이었습니다. 이를 위해 중립국시찰소조를 지정된 남북한 10개의 장소에 배치시켜 감독과 시찰을 실시하고 위반사건 발생시 특별감시와 시찰을 진행토록 하였습니다.*

중립국감독위원회 규모는 활동범위가 많았던 1953년 말에는 유엔군측 스위스와 스웨덴이 각각 100여 명, 공산군측 폴란드와 체코슬로바키아가 각각 300여 명 규모였으나 1956년 이후 기능과 활동범위가 감소됨에 따라 각 국가별 10여 명 수준으로 급감하였고, 현재는 유엔군측만 각각 5명씩으로 임무를 수행하고 있습니다.**

*정전협정 42조

**손규석 · 조성훈 · 김상원, 『6.25 전쟁과 유엔군』(서울 : 국방부 군사편찬연구소, 2015), 477쪽

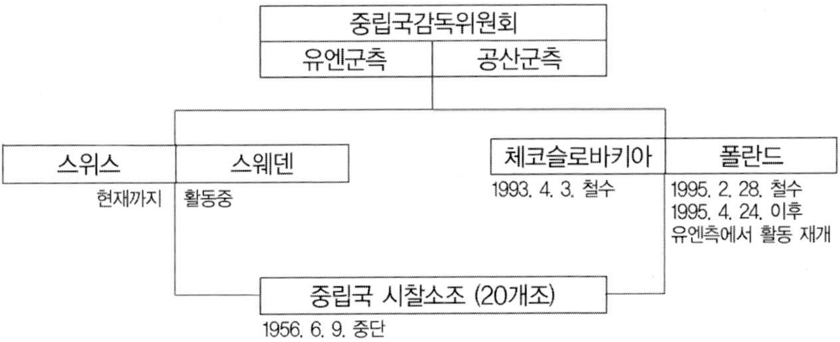

During the armistice negotiations, the UNC side proposed to bestow all post-war controls to the MAC composed of representatives of both parties, but the Communist side proposed to form a monitoring body made up of neutral representatives to control areas outside of the DMZ and the HRE because they didn't want the MAC to have jurisdiction over the entire Korean Peninsula, and thereby meddling with internal affairs of the other side. As UNC agreed to the North's proposal, the NNSC was formed(JSA Panmunjom, p. 195).

The NNSC was composed of four senior officers, two of whom were appointed by neutral nations nominated by the Commander of UNC, namely, Sweden and Switzerland, and two of whom were appointed by neutral nations nominated jointly by the Supreme Commander of KPA and the Commander of CPV, namely, Poland and Czechoslovakia(Paragraph 37, Armistice Agreement).

The NNSC operated 20 NNITs at the designated ports and airports of the other side to supervise and monitor any replacement or introduction of additional military personnel, combat equipment, weapons and ammunition onto the Korean Peninsula, which was prohibited, and reported the results to the MAC. Also, the NNSC conducted special investigations at places outside the DMZ for possible Armistice violations at the request of the MAC(Paragraph 41, Armistice Agreement).

Of the 20 NNITs, ten were permanently stationed and the other ten were held in reserve. An NNIT had four field grade officers, two from each side. The ten locations where the NNITs were permanently stationed were Inchon [Incheon], Taegu [Daegu], Pusan [Busan], Kangnung [Gangneung], and Kunsan [Gunsan] on the UNC side and Sinuiju, Chongjin, Hungnam, Manpo, and Sinanju on the North Korean side(Paragraph 43, Armistice Agreement).

정전협정 협상과정에서 유엔군측은 전후 모든 통제를 양 당사자 대표로 구성된 군사정전위원회에 일임하고자 제안했으나, 공산군측은 군사정전위원회가 한반도 전체를 관할함으로써 각 당사자 내부 문제까지 관여하는 것을 원치 않아 비무장지대와 한강하구지역 이외의 지역을 관할하는 중립국 대표들로 이루어진 감시조직을 두자는 제안을 하였고, 유엔군측이 동의하여 중립국감독위원회가 편성되었습니다.*

중립국감독위원회는 4명의 고급장교로 구성하되 그 중의 2명은 국제연합군 총사령관이 지명한 중립국 스웨덴과 스위스가 임명하며, 그 중의 2명은 조선인민군 최고사령관과 중국인민지원군 사령관이 공동으로 지명한 중립국, 즉 폴란드 및 체코슬로바키아가 이를 임명하는 것이었습니다.**

중립국감독위원회는 20개의 중립국시찰소조를 운영하여 상대방의 지정된 항만과 공항에 파견하여 한반도 외부로부터 증원되는 군사인원, 전투장비, 무기, 탄약의 교체 및 반입 금지에 대해 감시 · 감독하고 그 결과를 군사정전위회에 보고토록 하였습니다. 또한 군사정전위원회의 요청에 따라 비무장지대 이외 지역에서의 정전협정 위반사항을 조사하는 임무를 수행토록 하였습니다.***

20개 중립국시찰소조 중 10개는 상주하고 10개 조는 예비로 구성되며 각조는 4명의 영관급장교로 쌍방이 각각 2명씩 참석토록 하였습니다. 10개의 상주지역은 유엔사측은 인천, 대구, 부산, 강릉, 군산, 북한측 지역에는 신의주, 청진, 홍남, 만포, 신안주로 지정되었습니다.****

* 이문항, 『JSA-판문점(1953~1994)』(서울 : 소하, 2001), 195쪽

** 정전협정 37조

*** 정전협정 41항

**** 정전협정 43항

Paragraph 13.d of the Armistice Agreement states that the two sides should cease the introduction of new military equipment and materials to Korea so as to maintain the balance of power while allowing the replacement of destroyed or damaged items on the piece-for-piece basis only though the ten ports and airports of entry where the NNITs are stationed(JSA Panmunjom, p. 197).

After the Armistice Agreement was signed, the NNITs designated by the Communist side engaged in activities unrelated to the Armistice Agreement including espionage in the South and blindly supporting the North while deliberately interfering with activities of the NNITs of the UNC side in North Korea by delaying their investigations or refusing to investigate violations by the North(JSA Panmunjom, p. 198).

In addition, North Korea not only surreptitiously imported large quantities of military equipment such as fighters and tanks, but also interfered with the reporting activities by the NNSC of the UNC side.

In May 1956, at the 70th MAC meeting, the UNC side announced it would temporarily suspend all activities of the 20 NNITs. All NNITs withdrew from South Korea to the NNSC Headquarters in the DMZ by June 9, 1956 and from North Korea by June 12, 1956(JSA Panmunjom, p. 203).

At the 75th MAC meeting in June 1957, UNC pointed out that North Korea had violated Paragraph 13, Article 2 of the Armistice Agreement, which prohibits introduction of weapons to Korea, and announced that it would do away with the paragraph until North Korea began complying with the provision. North Korea, on the other hand, continued to condemn UNC for paralyzing the NNSC(JSA Panmunjom, p. 203).

정전협정 13항 'ㄹ' 목은 휴전기간중 남측과 북측이 세력균형을 유지토록 한국 외부로부터 새로이 군사장비 및 물자 도입을 중지하고 기간중 파괴, 파손 등의 물건은 1:1로 교환할 수 있도록 하였고, 이 또한 위에 언급된 중립국시찰소조가 상주하고 있는 10개의 항만과 공항만 활용토록 하였습니다.*

정전협정 체결 이후 공산군측 중립국시찰소조 요원들은 남한지역에서 정전협정과 무관한 간첩행위와 같은 일들을 자행하면서도 항상 북한측을 절대로 지지하였고, 유엔사측 중립국 시찰소조의 조사를 지연시키거나 북한측의 위반행위 조사를 자주 거부하는 등 북한지역 내 활동을 고의적으로 방해하였습니다.**

또한 북한은 다량의 전투기, 전차 등의 군사 장비를 몰래 반입할 뿐만 아니라 이를 보고하는 남측 중립국감독위원회의 활동을 방해하였습니다.

이에 1956년 5월 제70차 군사정전위원회 본회의에서 유엔군측은 20개 중립국시찰소조 활동을 잠정 중지할 것을 선언하였고, 1956년 6월 9일에 중립국감독위원회 시찰소조들이 모두 남한에서 DMZ 안에 있는 중립국감독위원회 본부로 철수했고 북한으로부터는 6월 12일까지 모두 철수했습니다.***

유엔군측은 1957년 6월 제75차 군사정전위원회 본회의에서 북한의 위반행위를 지적하고 한국으로의 무기반입을 금지한 정전협정 제2조 13항 관련조항 의무이행을 북한이 준수할 때까지 남측도 벗어날 것임을 선언하였습니다. 그러나 북측은 오히려 유엔사측이 중립국감독위원회 기능을 마비시켰다고 계속 비난하였습니다.****

* 이문항, 『JSA-판문점(1953~1994)』(서울 : 소하, 2001), 197쪽

** 이문항, 『JSA-판문점(1953~1994)』(서울 : 소하, 2001), 198쪽

*** 이문항, 『JSA-판문점(1953~1994)』(서울 : 소하, 2001), 203쪽

**** 이문항, 『JSA-판문점(1953~1994)』(서울 : 소하, 2001), 203쪽

On April 10, 1991, North Korea stated that the NNSC had made no tangible contribution whatsoever to resolving Armistice issues on the Korean Peninsula, and therefore had no reason to exist anymore. It was because the NNSC failed to prevent the appointment of a South Korean general officer as the Senior Member of UNCMAC. Since then, North Korea has continued to argue that the NNSC has no use. After the split of Czechoslovakia into the Czech Republic and Slovakia in April 1993, North Korea refused to acknowledge Czech as an NNSC delegation and denied their entry into the country the Czech Republic wanted to succeed the NNSC mission. In February 1995, North Korea forcefully deported the Polish delegation of the NNSC as well. On May 3, 1995, North Korea sent a message to UNC stating, "The office of the NNSC in Panmunjom has been shut down and the members of the NNSC are now banned from entering the northern side of the Joint Security Area without prior permission" (Military Armistice Commission Yearbook Vol. 8, p. 119).

Although the NNSC activities on the Communist side came to a halt, the Swiss and Swedish delegations of the NNSC on the UNC side have continued to operate since 1953. Furthermore, the Polish delegation, which was forced to leave North Korea, visits South Korea on a quarterly basis to attend NNSC meetings on the UNC side as per its policy to continue to honor its NNSC mission.

With the withdrawal of the NNSC on the Communist side in 1995, the NNSC became unable to execute its original mission. Therefore, the Commander of UNC proposed expanded tasks to the NNSC, which were agreed by the ROK, Switzerland, Sweden and UNC.

한국군 장성의 유엔군사령부 군사정전위원회 수석대표 임명 이후 북측은 중립국감독위원회가 한국군 수석대표 임명을 사전에 막지 못했다는 이유로 1991년 4월 10일 중립국감독위원회가 실질적으로 한반도의 정전협정 문제 해결에 기여하는 바가 조금도 없으며, 이제 중립국감독위원회는 더 이상 존재할 필요가 없다고 언급한 이후 중립국감독위원회 무용론을 지속 주장하였습니다. 그리고 1993년 4월, 체코슬로바키아가 체코와 슬로바키아로 분리 독립하면서 체코가 계승하여 임무수행을 하고자 하였으나 북한은 입국자체를 거부함으로써 철수시켰으며, 1995년 2월 폴란드 대표단마저 중립국감독위원회에서 강제 철수시켰습니다. 1995년 5월 3일 북한은 "판문점 중립국감독위원회 사무실을 폐쇄하고 중립국감독위원회 요원들은 사전 허가 없이 공동경비구역 북측 지역의 출입을 제한한다"는 것을 유엔군측에 통보하였습니다.*

*합참정보본부, 『군사정전위원회 편람』(제8집), 119쪽

공산군측 중립국감독위원회 활동은 정지되었으나 유엔군측 중립국감독위원회 스웨덴과 스위스 대표단은 현재까지 운용되고 있으며, 공산군측 폴란드 대표단은 본국으로 철수하였지만 중립국감독위원회 임무를 계속 수행한다는 원칙을 세우고 현재까지 매분기 방한하여 유엔군측 중립국감독위원회 회의에 참여해 오고 있습니다.

공산군측 중립국감독위원회 철수로 1995년 이후 최초 부여된 목적의 임무수행이 제한됨에 따라 유엔군 사령관은 중립국감독위원회의 추가 과업을 제안하였으며, 한국, 스위스, 스웨덴, 유엔사가 확대과업에 합의하였습니다.

• 중립국감독위원회의 확대 과업은 2005, 2010, 2016년도에 합의되었습니다.

NNSC's Expanded Tasks Proposed by CDR, UNC

1. Observe UNCMAC's Armistice operations
2. Observe and augment UNCMAC's Armistice education and orientation program
3. Observe military exercises
4. Observe UNC/USFK force deployment activities

The NNSC continues to contribute to implementing the Armistice Agreement and maintaining stability on the Korean Peninsula by actively participating in such activities as UNCMAC's investigations into KPA's Armistice violations including the sinking of the ROKS Cheonan and the shelling of Yeonpyeong Island, observation of ROK-U.S. combined and ROK-only exercises, and verification of CMA implementation including GP withdrawal and mine removal at the Joint Security Area.

Prisoners of War

The issue of repatriating POWs was a point of such sharp contention that it took the two sides two years to reach a tentative agreement. The discussion on the issue began at the 44th meeting and the deal wasn't struck until the 144th meeting. Another factor that made the issue so complicated was the fact that many Communist POWs refused to return to their sides. In the process, a significant number of the ROK POWs didn't get repatriated to South Korea (The Korean War and UN

유엔군 사령관 제안 중립국감독위원회
확대 과업내용

1. 유엔사 군정위의 정전협정관련 작전활동 참관
2. 유엔사 군정위의 정전협정교육 및 견학지원
3. 군사연습 참관
4. 유엔사 및 주한미군사 병력 전개 및 활동 참관

GP 철수, 철거 현장검증
Participation in On-site Verification of GP Withdrawal

최근에는 천안함 피격사건, 연평도 포격도발 등 북한군의 정전협정 위반사건을 조사하는 유엔사 군사정전위원회 활동에 함께하였고 한 · 미 연합 및 한국군 단독훈련에 참관, GP철수 및 JSA 지뢰제거시 검증활동으로 군사분야합의서 이행에 참여하는 등 적극적인 활동으로 정전협정 이행과 한반도 안정 유지에 기여하고 있습니다.

전쟁포로에 관한 조치

전쟁포로의 송환문제는 44차 회의에서 시작하여 144차 회의에서 잠정합의에 이를 정도로 정전회담 2년 내내 첨예하게 입장대립을 보였습니다. 전쟁포로 송환에 쉽게 합의하지 못한 또 다른 이유는 송환을 거부하는 반공포로가 많았기 때문이었습니다. 이 과정에서 한국군 포로들은 상당부분 북한에서 돌아오지 못하였습니다.* 실질적으로 포로로 잡힌 인원들 중에는 남한 젊은이들이

*손규석 · 조성훈 · 김상원, 『6.25 전쟁과 유엔군』(서울 : 국방부 군사편찬연구소, 2015), 448쪽

Forces, p. 448). Recognizing the complexity of POW composition such as young South Koreans forced to serve in the KPA, the UNC insisted on giving POWs the freedom to choose, but the Communist side demanded unconditional repatriations of all POWs(The Korean Peninsula Armistice Regime, p. 159).

In this stalemate, the International Committee of the Red Cross demanded immediate repatriation of sick and injured POWs in December 1952. The Communist side agreed to UNC's proposal to immediately repatriate sick and injured POWs on March 1953, and as a result, UNC handed over 6,670 POWs and the Communist side 684 POWs between April and May 1953(The Korean Peninsula Armistice Agreement, p. 160).

The two sides agreed to complete the repatriation of all POWs directly to the side to which they belonged at the time of capture within 60 days after the Armistice Agreement became effective. To expedite the process, the two sides agreed to exchange the total number of POWs by nationalities and each group of POWs delivered to the other side to be accompanied by rosters to include name, rank and internment or military serial number. All the remaining POWs that were not directly repatriated were handed over to the Neutral Nations Repatriation Commission (NNRC) (Paragraph 51, Armistice Agreement).

인민군으로 강제 징용된 경우 등 다양한 인원으로 구성된 것을 파악한 유엔군측은 "포로송환은 자유의사에 맡기자"고 제안하였으나, 공산군측은 "포로전원의 무조건 송환"을 요구하였습니다.*

이러한 교착상태에서 1952년 12월 국제적십자위원회는 부상당하거나 병이 든 상병포로의 즉각 송환조치를 요구하였습니다. 유엔군측 상병포로의 즉각 송환 제의에 공산군측이 1953년 3월 동의하였고, 이에 따라 4월에서 5월까지 유엔군측은 6,670명, 공산군측은 684명의 포로를 인계하였습니다.**

포로송환협의 결과 정전협정이 효력을 발생한 후 60일 이내에 양측은 전체 전쟁포로를 포로가 된 당시에 그들이 속한 일방에 직접 송환하고 이를 위해 국적별로 분류한 총수를 교환하며 각 집단은 국적별로 작성한 명부를 휴대하되 이에는 성명, 계급 및 수용번호 또는 군번호를 포함하도록 하였습니다. 직접 송환하지 않은 나머지 전쟁포로들 모두는 중립국송환위원회에 넘겨 처리토록 하였습니다.***

* 이상철, 『한반도 정전체제』(서울 : 한국국방연구원, 2012), 159쪽

** 이상철, 『한반도 정전체제』(서울 : 한국국방연구원, 2012), 160쪽

*** 정전협정 51항

⟨Agreement on Repatriation of Prisoners of War(June 8, 1953)⟩

1. All POWs who wish to return to their homeland shall be repatriated within 60 days.
2. No force shall be used against POWs to effect or prevent their repatriation.
3. The custody of POWs after 60 days shall be transferred to the Neutral Nations Repatriation Commission.
4. Within 90 days after the Neutral Nations Repatriation Commission takes over the custody, the nations to which POWs belong shall send representatives to explain to them their rights to return home.
5. All explanations shall be conducted in the presence of a representative of the Neutral nations Repatriation Commission and a representative from the detaining side.
6. At the expiration of 90 days after the transfer of the custody, the question of disposition of the POWs who have not exercised their right to be repatriated shall be submitted to the Political Conference and shall be settled within 30 days.
7. The status of POWs who have not exercised their right to be repatriated within 120 days shall be declared civilian.
8. Those POWs who choose to go to neutral nations shall be assisted by the Neutral Nations Repatriation Commission.

The Committee for Repatriation of Prisoners of War was composed of six field grade officers, three from each side. Under the general supervision and direction of the MAC, the repatriation committee was

〈포로 송환 협정(1953년 6월 8일)〉

1. 본국으로 복귀를 원하는 모든 전쟁포로는 60일 이내에 송환
2. 송환에 영향을 미치거나 송환을 막기 위해 힘에 의한 위법의 사용금지
3. 60일 이후 나머지 포로는 중립국송환위원회에 인도
4. 그 이후 90일 동안 포로 소속국가의 파견대표가 본국으로 복귀될 수 있는 권리에 대한 설명을 실시
5. 설명은 중립국송환위원회 대표와 억류측 대표의 면전에서 실시
6. 90일 기한이 마감되면 송환거부 포로는 정치회담으로 인도되며, 30일 이내에 결정을 내림
7. 그 기간 이후에도 남는 송환 거부 포로는 민간인 신분으로 규정
8. 만일 그들이 중립국으로 가기를 희망한다면 중립국송환위원회가 후원

전쟁포로송환위원회는 양측의 영관급장교 3명씩 총 6명으로 구성하여 군사정전위원회의 감독과 지도하에서 책임지고 전쟁포로 송환의 임무를 수행토록 하였으며,

responsible for conducting the repatriations of POWs and ensuring their welfare though the Joint Red Cross Teams. When unable to reach agreement on any matter relating to its responsibilities, the committee was to immediately refer such matter to the MAC for decision (Paragraph 56, Armistice Agreement).

At that time, the UNC repatriated a total of 82,493 POWs (75,823 KPA POWs and 6,670 Chinese POWs) who wished to return and were sick and injured, and the Communist side repatriated a total of 13,457 POWs (8,333 ROK POWs and 5,124 UNC POWs). Although North Korea officially announced the number of ROK and UNC POWs to be 108,257 on June 25, 1951, there were only 11,559 POWs on the list it exchanged at the armistice negotiation table in December the same year. Furthermore, according to the data that UNC submitted to the UN, the number of POWs and missing persons was 82,318. Given that the number of ROK POWs repatriated by the Communist side was only 8,333, presumably, a significant number of ROK POWs were detained in the North and have never returned to the South. However, North Korea maintains its position that there is not a single ROK POW in North Korea (The Korean War and UN Forces, p. 452).

To ensure all POWs have the opportunity to exercise their right to be repatriated, Sweden, Switzerland, Poland, Czechoslovakia and India were requested by both sides to appoint a member each to the NNRC (The Korean War and UN Forces, p. 183). The NNRC was responsible for repatriating POWs who wished to return home and controlling those POWs who refused to be repatriated(Paragraph 51, Armistice Agreement). Upon completion of its mission, the NNRC was disbanded in 1954.

이를 지원하기 위하여 공동적십자 소조를 설립하여 전쟁포로의 복지를 담당토록 하였습니다. 임무수행간 관련 의제가 쌍방 합의에 도달하지 못하면 즉시 군사정전위원회에 제기하여 결정토록 하였습니다.*

정전협정 체결 당시 송환희망 포로와 상병포로 대상 유엔군측은 총 82,493명(북한군 75,823명, 중공군 6,670명)의 공산군 포로를 송환하였고, 공산군측은 총 13,457명(한국군 8,333명, 유엔군 5,124명)의 포로를 유엔군측에 송환하였습니다. 1951년 6월 25일 북측은 국군과 유엔군 포로 규모를 108,257명이라고 공식발표하였으나, 같은 해 12월 휴전협상 포로명단 교부시 11,559명에 불과한 명단을 제공하였습니다. 그리고 1953년 8월 7일 유엔군사령부가 유엔에 제출한 당시 국군포로 및 실종자 수는 82,318명으로 집계하였고, 공산군측이 송환한 국군출신 포로수가 8,333명에 불과한 사실을 보면 상당수의 국군포로가 송환되지 못하고 북한에 억류당한 것으로 추정이 됩니다. 그러나 북한은 단 한 명의 국군포로도 북측에 존재하지 않는다는 기본입장을 고수하고 있습니다.**

중립국송환위원회는 전체 전쟁포로로 하여금 정전 후 피송환권 행사의 기회를 가지도록 하기 위하여 쌍방이 스위스, 스웨덴, 폴란드, 체코슬로바키아 및 인도에 각각 1명씩 위원을 임명하도록 요청하여 설립되었고*** 송환희망 포로 및 전쟁 이전 자기 나라로 돌아가지 않겠다는 송환거부 포로에 대한 관리 및 송환임무를 직접 지도하였습니다.**** 이후 1954년 임무를 완수하고 해체가 되었습니다.

* 정전협정 56항

** 손규석 · 조성훈 · 김상원, 『6.25 전쟁과 유엔군』(서울 : 국방부 군사편찬연구소, 2015), 452쪽

*** 손규석 · 조성훈 · 김상원, 『6.25 전쟁과 유엔군』(서울 : 국방부 군사편찬연구소, 2015), 183쪽

**** 정전협정 51항

Recommendation to Governments Concerned on Both Sides

The military commanders of both sides recommended to the governments concerned that, within three months after the Armistice Agreement was signed and became effective, a political conference of a higher level be held by representatives of both sides to settle through negotiation the questions of the withdrawal of all foreign forces from Korea, the peaceful settlement of the Korean question, etc. (Paragraph 60, Armistice Agreement).

Following the Berlin Conference on February 28, 1954, which was a quadrilateral meeting among the foreign ministers of the U.S., the U.K., France and Soviet Union, the Geneva Conference was held between April 26 and June 15, 1954. The foreign ministers of the 15 countries that participated in the Korean War (minus South Africa) along with China, Soviet Union and the two Koreas discussed a peaceful resolution to the Korean problem. UNC emphasized the role and functions of the UN on the Korean Peninsula whereas the Communist side demanded the withdrawal of UNC.

On the surface, the political talks to find a peaceful resolution to the Korean problem appeared to have some common agenda including establishing a parliament for free elections, international surveillance, and withdrawal of foreign troops. However, the UNC's position was to establish a unified Korean government with a free election under UN surveillance whereas the Communist side insisted on dealing with the Korean problem outside the UN, claiming all UN resolutions on the Korean Peninsula had been illegal and invalid. The political talks were bound to break down as the North demanded the withdrawal of the U.S. forces and opposed to the signing of a mutual defense treaty with the U.S. while refusing to have a free election on the Korean

쌍방 관계정부들에 건의

쌍방 군사령관은 쌍방의 관계 각국 정부에 정전협정이 조인되고 효력을 발생한 후 3개월 내에 각기 대표를 파견하여 쌍방의 한 급 높은 정치회의를 소집하여 외국군대의 철거 및 한국문제의 평화적 해결 등을 협의할 것을 건의토록 하였습니다.*

*정전협정 60항

1954년 2월 28일 미 · 영 · 불 · 소 4개국 외무부장관간의 베를린회담을 거쳐 한국전쟁에 참전했던 15개국(남아공 불참), 남 · 북한, 중국, 소련 등 19개국 외무장관들이 참석한 제네바 정치회의가 1954년 4월 26일부터 6월 15일까지 한국 문제의 평화적 해결을 논의하기 위하여 개최되었으나 유엔군측은 한반도에서 유엔의 역할과 활동을 강조한 반면 공산군측은 유엔사의 철수를 주장하였습니다.

스위스 제네바 1954년 정치회담장
At the venue of the 1954 Geneva Conference in Switzerland

한반도 문제의 평화적 해결을 위한 정치회담은 표면적으로는 자유선거를 위한 의회구성과 국제감시, 외국군 철수 등의 공통적인 요소를 포함하고 있었으나, 유엔군측은 유엔감시하에 자유선거로 통일한국 정부를 수립해야 한다는 입장이었고, 공산군측은 한반도 문제 관련 유엔의 모든 결의안이 불법적이고 무효이므로 유엔 밖에서 한국문제를 해결해야 한다는 입장이었습니다. 당시 남측 인구가 2,000만여 명이 넘는데 북한의 경우는 800만여 명에 불과해서 북측은 선거방식에 동의하려 하지 않았고, 미군 철수와 한미상호방위조약 체결 반대 등을 내세워 정치회담은 결렬될 수밖에 없었습니다.**

**손규석 · 조성훈 · 김상원, 『6.25 전쟁과 유엔군』(서울 : 국방부 군사편찬연구소, 2015), 489쪽

Peninsula. At that time, the South had a population of more than 20 million whereas the North only had 8 million people(The Korean War and UN Forces, p. 489).

In the Declaration by the Sixteen on June 15, 1954, the 16 UN Sending States proclaimed, "The UN, under its Charter, is fully and rightfully empowered to take collective action to repel aggression, to restore peace and security, and to extend its good offices to seeking a peaceful settlement in Korea. In order to establish a unified, independent and democratic Korea, genuinely free elections should be held under UN supervision, for representatives in the National Assembly, in which representation shall be in direct proportion to the indigenous population in Korea" (The Korean War and UN Forces, p. 490).

In the end, the Political Conference on Korea in Geneva ended after 87 days without any tangible results other than a joint declaration that the Korean issue should be transferred to the UN (The Korean War and UN Forces, p. 488). Since then, we have continued to pursue negotiations for peace, such as high-level inter-Korean talks, the four-party talks, the six-party talks, and the inter-Korean summit talks.

United Nations Command

The United Nations Command, a U.S.-led unified command under the UN Security Council resolution dated July 7, 1950, has the legal mandate to implement the Armistice Agreement as a direct party to the Korean War and a signatory of the Armistice Agreement(The Korean Peninsula Armistice Regime, p. 91).

The UNC was relocated from Tokyo to Seoul on July 1, 1957, to

유엔 참전 16개국은 1954년 6월 15일 "유엔은 그 헌장에 의거하여 침략을 격퇴하기 위하여 공동작전을 하고 평화와 안전을 회복하여 한국에서 평화적인 해결을 추구하는 높은 과업을 확장할 권리를 충분히 가지고 있다는 점을 재확인하고, 통일된 민주주의적 독립한국을 건설하기 위하여 유엔감시하에 인구비례에 따라 대표자가 선출되는 국회의원의 진정한 자유선거를 한국에서 시행한다"는 공동선언을 채택하였습니다.*

**손규석 · 조성훈 · 김상원, 『6.25 전쟁과 유엔군』(서울 : 국방부 군사편찬연구소, 2015), 490쪽

결국 유엔군측은 한반도 문제는 유엔으로 이관할 수밖에 없다는 요지의 공동선언을 발표하고 87일간의 제네바 정치회담은 성과 없이 종료되었습니다.** 이후에도 남북고위급 회담, 4자회담, 6자회담, 남북정상회담 등 지속적으로 평화를 위한 협상노력을 지속해 오고 있습니다.

**손규석 · 조성훈 · 김상원, 『6.25 전쟁과 유엔군』(서울 : 국방부 군사편찬연구소, 2015), 488쪽

유엔군사령부

유엔군사령부는 1950년 7월 7일 유엔안보리 결의안에 의거 미국 주도하의 통합군사령부로 창설되었으며, 6.25 전쟁의 수행자이고 정전협정의 서명자로서 정전협정 이행에 대한 법적 지위를 유지하고 있습니다.***

***이상철, 『한반도 정전체제』(서울 : 한국국방연구원, 2012), 91쪽

정전 이후 1957년 7월 1일 도쿄에서 서울로 유엔군사

exercise operational control over the ROK armed forces, which was later transferred to the ROK-U.S. Combined Forces Command (CFC) established on November 7, 1978, and to manage the Armistice regime. UNC and CFC maintain a coordination relationship, but when it comes to Armistice issues, UNC exercises control over CFC. UNC is directly commanded by the U.S. Joint Staff(Military Armistice Commission Yearbook Vol. 8, p. 3).

UNC Organization Chart

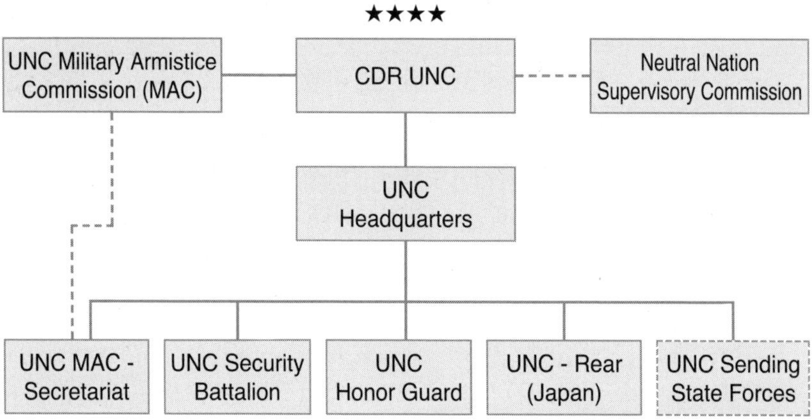

UNC also serves as a force provider by providing control and support for Sending State forces during a conflict on the Korean Peninsula and maintaining the UNC Rear bases in Japan.

The UNC was established at Camp Zama in Tokyo in July 1951 where the U.S. Far East Command was located. When the U.S. and

령부를 이전하여 한국군에 대한 작전통제 및 정전체제를 관리하였으며, 1978년 11월 7일 한미연합군사령부 창설 후 한국군에 대한 작전통제권을 한미연합군사령부로 전환하였습니다. 유엔군사령부와 한미연합군사령부는 협조관계이나 정전협정업무에 관한 사항은 유엔군사령부가 한미연합군사령부를 통제하여 업무를 수행하며 유엔군사령부는 미 합참으로부터 직접 지휘를 받고 있습니다.*

*합참정보본부, 『군사정전위원회 편람』(제8집), 3쪽

유엔사 조직도

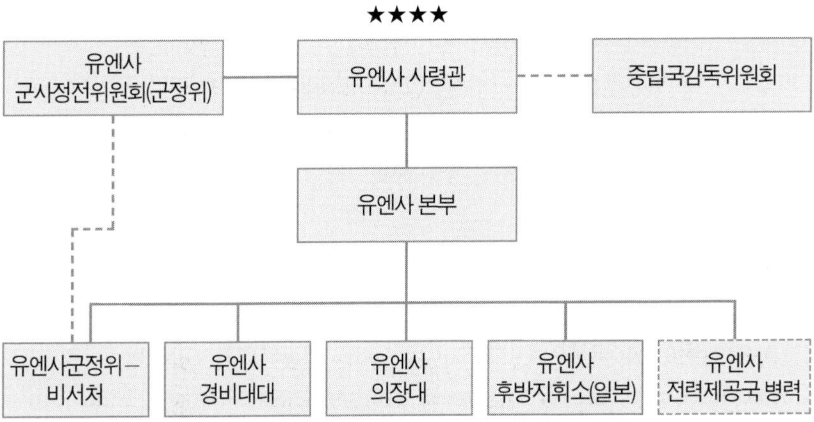

또한, 유엔군사령부는 한반도 전쟁 재발시 파견되는 유엔회원국 군대에 대한 통제 및 지원, 주일 유엔사 후방기지의 유지 등 한반도에 전력 제공자로서 임무를 수행하고 있습니다.

유엔군사령부는 1951년 7월 미극동사령부가 위치한 일본 동경의 자마기지에 창설되었으며 1951년 9월 미국

Japan signed a peace treaty on September 1951, the two countries agreed to make some U.S. bases in Japan available for UN forces participating in the Korean War for logistical support. The UNC and the government of Japan (GOJ) signed the UNC-GOJ Status of Forces Agreement (SOFA) in February 1954 to ensure those bases would remain available even after the Armistice Agreement. Originally, eleven countries signed up for the UNC-GOJ SOFA, namely the U.S., the U.K., Canada, France, Australia, the Philippines, New Zealand, Italy, Turkey, Thailand and South Africa, and currently nine countries, all the original members minus Italy and South Africa, remain committed to the agreement(The Korean Peninsula Armistice Regime, p. 98).

In July 1957, with the UNC's relocation from Tokyo in Japan, to Seoul in South Korea, UNC established the UNC Rear in Japan to maintain the UNC-GOJ SOFA. UNC Rear was originally located at Camp Zama, but moved to Yokota Air Base in November 2007, where it continues serve as a rear command post of UNC.

Currently, eight countries, the U.K., Australia, New Zealand, Canada, France, the Philippines, Thailand, and Turkey, maintain a liaison officer to the UNC Rear(The Korean Peninsula Armistice Regime, p. 99). There are seven UNC Rear bases in Japan.

과 일본은 평화조약을 체결할 당시 한국전에 참전한 유엔군에 대한 군수지원을 위해 일본내 미군기지를 사용하기로 합의하였습니다. 정전협정 체결 이후에도 이를 보장할 수 있도록 1954년 2월 유엔사와 일본 정부간에 유엔군 주둔군지위협정(SOFA)을 체결하였습니다. SOFA 체결 당시 미국, 영국, 캐나다, 프랑스, 호주, 필리핀, 뉴질랜드, 이탈리아, 터키, 태국, 남아공 11개국이 원년 회원국으로 참여하였고, 현재 회원국은 이탈리아와 남아공이 빠진 9개국입니다.*

*이상철, 『한반도 정전체제』(서울 : 한국국방연구원, 2012), 98쪽

1957년 7월 유엔사가 일본 동경에서 한국 서울로 이동함에 따라 유엔사와 일본간의 SOFA 협정을 유지하기 위하여 유엔사 후방지휘소를 창설하였습니다. 최초에는 자마 기지에서 주둔해 오다가 2007년 11월 요코타 기지로 이전하여 현재까지 유엔군사령부의 후방지휘소 임무를 수행하고 있습니다.

현재 유엔사 후방지휘소에 연락장교를 파견한 국가는 영국, 호주, 뉴질랜드, 캐나다, 프랑스, 필리핀, 태국, 터키 등 8개국입니다.** 일본 내에는 7개의 유엔사 후방기지가 지정되어 있습니다.

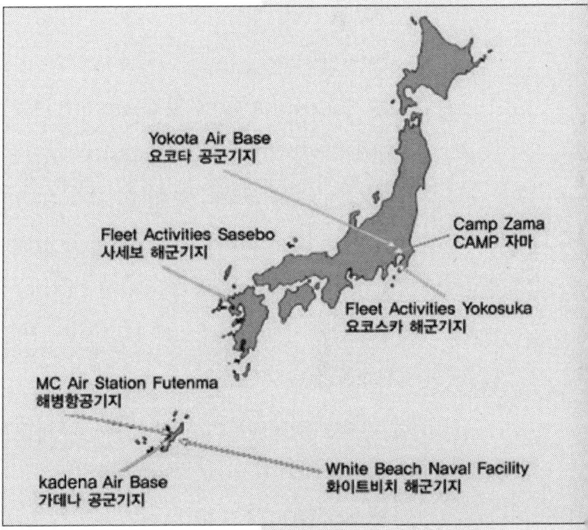

일본에 위치하고 있는 유엔군사령부 육군, 해군, 공군, 해병대 후방기지/ The Army, Navy, Air Force and Marine Bases of UNC Rear in Japan

**이상철, 『한반도 정전체제』(서울 : 한국국방연구원, 2012), 99쪽

〈UNC Rear Command and Bases〉*

Based on the UNC-GOJ SOFA, UNC Sending States forces have free access to the UNC Rear bases by notification only without prior approval of the GOJ in the event of a contingency on the Korean Peninsula.

- Yokota Airbase is home to UNC Rear Command and U.S. Forces Japan(USFJ) HQ.
- Camp Zama is home to the U.S. Army Japan.
- Fleet Activities Yokosuka is home to the U.S. Fleet and the U.S. Naval Forces Japan. The command ship of the Fleet, is homeported at Yokosuka as well as an aircraft carrier, cruisers, destroyers, mine clearing ships, maintenance ships, and supply ships.
- Fleet Activities Sasebo supports the U.S. Fleet and is home to several tenant units including an amphibious squadron. It also stores the largest amount of fuel and ammunition in the Western Pacific.
- Kadena Airbase is home to the Wing equipped with reconnaissance, strike and air-refueling capabilities. Fighter jets can reach the Korean Peninsula in 0 hour and cargo aircraft in 0 hours from Kadena.
- Marine Corps Air Station Futenma provides airlift to the U.S. marines for deployments to the Korean Peninsula. It is capable of accommodating all types of U.S. air assets with a long runway.
- White Beach Naval Facility is a staging area for Marines and their equipment.

〈유엔군사령부 후방지휘소 및 기지〉*

유엔군사령부와 일본정부간 SOFA에 따라 한반도 유사시 유엔군사령부 회원국 전력이 일본정부의 사전 허가절차 없이 통보만으로 유엔군사령부 후방기지의 자유로운 사용이 가능합니다.

- 요코타 공군기지는 유엔군사령부 후방지휘소 본부와 주일미군사령부가 위치
- 자마 육군기지는 주일 미육군사령부가 위치
- 요코스카 해군기지는 함대 사령부와 주일 미 해군사령부가 위치하고 있으며 지휘함과 항모, 순양함, 구축함, 소해함, 정비함, 보급함 등으로 구성

- 사세보 해군기지는 함대를 지원하고 상륙전대 등 여러 부대가 주둔하고 있으며 서태평양 지역 최대의 유류 및 탄약을 저장

- 가데나 공군기지는 전투비행단 등 정찰 및 타격, 공중급유 능력을 보유하고 있으며 한반도까지는 전투기 0시간, 수송기 0시간 이내 도달 가능함.

- 후텐마 해병대 기지는 미 해병의 한반도 전개시 항공수송을 지원하고 모든 미군 항공자산의 이착륙이 가능한 활주로를 보유하고 있음.
- 화이트 비치 해군기지는 해병의 탑재부두임

*Military Armistice Commission Yearbook Vol. 8, p. 8/ 합참정보본부, 『군사정전위원회 편람』(제8집), 8쪽

요코타 공군기지

CAMP 자마

요코스카 해군기지

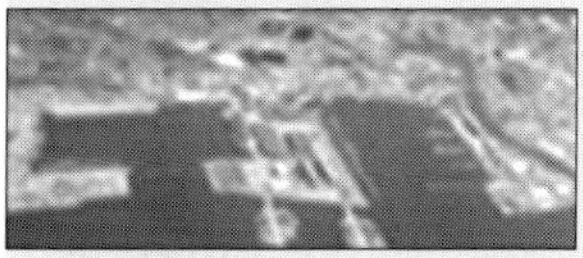
사세보 해군기지

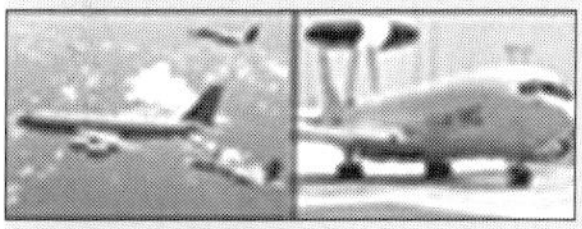
가데나 공군기지

후텐마 해병항공기지

화이트비치 해군기지

Note: After the Armistice Agreement was signed, the 16 UN Sending States proclaimed at the UN HQ their commitment to providing prompt support for the Republic of Korea if another war breaks out on the Korean Peninsula.

The UNC Rear bases where the forces under the U.S. Indo-Pacific Command are stationed provide deterrence against war on the Korean Peninsula in armistice and serve as logistics hubs through which USFJ and UNC Sending States forces will flow into the Korean Peninsula in contingencies(The Korean Peninsula Armistice Regime, p. 100).

The UNC Liaison Group was initially set up to coordinate policies and procedures for integrating forces from the Sending States during the Korean War. After the signing of the Armistice Agreement, defense attachés or diplomats at the Sending State embassies were accredited as UNC liaison officers. They have been involved in various activities including serving as the Rotating Member of UNCMAC, providing advice on Armistice-related issues, and participating in special investigations into Armistice violations as an investigator. Currently, the group is referred to as the UNC Sending State Liaison Group and is made up of 18 countries including the U.S. and the ROK.

The ROK Advisory Group to UNCMAC began its operation on December 20, 1953 when a ROK military MAC liaison officer was sent to the Ministry of Foreign Affairs to provide advice and liaison on Armistice issues. The group was later named as the ROK Military Group in 1957 and was put under the MND in 1972. It received its current title, the ROK Advisory Group to UNCMAC (on April 12, 1999 per MND Directive No. 626). Its mission includes providing assistance

참고: 정전협정이 체결된 이후 16개 유엔참전국들은 전쟁 재발시 신속히 한국을 지원할 것 1953년 7월 27일에 유엔본부에서 결의를 하였습니다.

미 인도태평양사령부 예하 전력이 배치되어 있는 유엔사 후방기지는 평시 한반도 전쟁억제의 역할을, 그리고 한반도 유사시에는 주일미군 전력의 즉각 출동 및 병참기지 역할, 그리고 유엔사 회원국의 전력투입을 위한 기지로서의 역할을 수행하고 있습니다.*

유엔군사령부 참전국 연락단은 6.25한국전쟁 당시 참전국의 군대를 유엔군사령부에 합류시키는 정책과 절차 문제 등 업무협조를 위해 설치되었습니다. 정전협정체결 이후 참전국 연락단의 구성인원은 대사관의 무관 또는 외교관들이 겸직하면서 군사정전위원회의 순환직 대표, 정전관련 문제에 대한 자문, 정전협정 위반에 대한 조사관 등 다양한 활동을 하고 있습니다. 현재 '유엔군사령부 회원국 연락단' 이라는 명칭으로 활동하고 있으며 미국, 한국 등 18개국입니다.

군사정전위원회 한국군 연락단은 정전업무의 자문과 연락임무를 위해 1953년 12월 20일 외무부에 한국군 군사정전위원회 연락장교를 파견하면서 시작되었습니다. 1957년에는 대한민국 군대표단으로 개명하였고, 1972년 국방부로 소속이 변경되었으며, 현재는 군사정전위원회 한국군 연락단(국방부 훈령 제626호, 1999년 4월 12일)으로 개칭되어 군사정전위원회의 한국대표를 보좌하면서

* 이상철, 『한반도 정전체제』(서울 : 한국국방연구원, 2012), 100쪽

• 유엔군사령부 참전국 연락단(18개국) : 호주, 벨기에, 캐나다, 콜롬비아, 덴마크, 프랑스, 그리스, 이탈리아, 네덜란드, 뉴질랜드, 노르웨이, 대한민국, 남아공, 필리핀, 태국, 터키, 영국, 미국

to the ROK Member of UNCMAC, liaises on UNC-related issues on behalf of the ROK military, attending various military and diplomatic meetings related to UNC, and supporting orientation and education programs in Panmunjom.

The NNSC on the UNC side, whose headquarters is located in PMJ, is composed of ten officers including the Swedish Member and the Swiss Member and supports the implementation of the Armistice Agreement and the activities of the MAC.

The UNC component of the MAC is made up of five members appointed by the UNC CDR including the Senior Member, ROK Member, U.K. Member, U.S. Member and Rotating Member. Its mission is to implement the Armistice Agreement and is supported by the Secretariat in accomplishing its mission.

The Secretariat of UNCAMC, which has a U.S. Colonel as the Secretary and a ROK Colonel as the Deputy Secretary, carries out dual functions; one as a special staff to the command and the other as an operational unit implementing the Armistice Agreement.

The main functions of the UNCMAC Secretariat as a special staff are 1) supporting the MAC, 2) providing advice on Armistice affairs for the UNC, ROK military, UNC Sending States and the NNSC, 3) supporting and coordinating with the UNCMAC Advisory Group, and 4) supporting and coordinating with the NNSC. The UNCMAC Secretariat continues to implement and monitor the Armistice Agreement by providing Armistice education to commanders and troops of the frontline units and military academic institutions including the Korean National Defense University (advanced security course) and Joint Forces Military University (O-5/O-6 command course) to raise awareness of the Armistice Agreement, and thereby reinforce compliance.

한국군의 대유엔군사령부 업무, 유엔군사령부 관련 각 종 군사외교 및 회의 참석, 판문점견학 지원 등 임무를 수행하고 있습니다.

유엔군측 중립국감독위원회는 스웨덴과 스위스 대표 등 총 10명으로 구성되어 판문점에 위치하고 있으며, 정전협정 이행과 군사정전위원회 활동을 지원하고 있습니다.

유엔사측 군사정전위원회는 유엔군사령관이 임명한 수석 대표, 한국군 대표, 영국군 대표, 미국군 대표, 순환 대표 등 총 5명으로 구성되어 정전협정이행 임무를 수행하고 있으며, 군사정전위원회 비서처가 군사정전위원회 임무수행을 지원하고 있습니다.

유엔군사령부 군사정전위원회 비서처는 미군 대령이 비서장으로 한국군 대령이 부비서장으로 보직되어 유엔군사령부의 특별참모부로서 참모기능과 정전협정을 이행하기 위한 실제 작전기능을 함께 수행하고 있습니다.

유엔군측 군사정전위원회 비서처 주요 참모기능은 첫째 군사정전위원회를 지원하고, 둘째 유엔군사령부, 한국군, 유엔사 회원국, 중립국감독위원회에 정전관련 자문, 셋째 군사정전위원회 자문단 협조 및 지원, 넷째 중립국감독위원회 협조 및 지원을 하고 있습니다. 또한 정전협정에 대한 이해를 증진시키고 정전협정 준수를 강화토록 전방사단 지휘관 및 장병, 국방대학교 안보과정, 합동대학교 지휘관리 과정 등을 대상으로 정전협정교육을 실시하는 등 정전협정 이행과 감독을 지속하고 있습니다.

정전협정교육
Armistice Education

The major operational activities of the UNCMAC Secretariat include 1) exchanging messages, holding meetings and conducting daily communication checks with the KPA, 2) maintaining 24/7 hotline with the KPA, 3) serving as a communication channel with the North on behalf of the Korea National Red Cross on humanitarian matters, and 4) managing access to the DMZ and crossing of the MDL. Also, it inspects frontline Army Divisions and education/orientation sites in the DMZ, and investigates violations of the Armistice Agreement such as the mine provocation in the 1st Division's area of operation using its special investigation team. In addition, it conducts a helicopter flight operation once a month called the JSA-Panmunjom H-128 Flight to exercise and validate its right to fly into Panmunjom and observes maritime live fire training in the Northwest Islands.

Also, the Commander of UNC hosts a Ambassadors roundtable to share current security environment, North Korean movements and key updates on UNC issues with the Ambassadors and the MAC Members. In addition, the Senior Member of UNCMAC hosts an UNCMAC Advisory Group meeting attended by the UNC Sending State Liaison Group and the ROK Advisory Group to share updates on special investigations, guard post inspections and other matters requiring cooperation. Furthermore, the Senior Member, together with the Chief of Staff of UNC, NNSC and UNCMAC Secretariat, meets with the NNSC in the forum called the UNC-NNSC Executive Committee to discuss NNSC operations and issues requiring cooperation and to guarantee conditions for the NNSC to continue to function.

GEN Walter L. Sharp, a former UNC Commander, stressed the importance of the MAC in the face of North Korea's provocations, aggressions and violations of the 1953 Armistice Agreement and

주요 작전활동은 첫째 북한군과의 통지문 발신과 수신, 회담 및 일일통신 점검실시, 둘째 유엔사와 북한군간 24시간 직통선 유지, 셋째 인도주의적 사안관련 대한적십자사의 대북창구역할 대행, 넷째 비무장지대 출입 및 군사분계선 통행관리 등입니다. 또한 비무장지대내 관련 사단 및 안보 견학장을 점검하고, 특별조사반과 GP/OP 점검단을 활용하여 1사단 지뢰도발 관련조사 등 정전협정 위반사항을 조사하였습니다. 그리고 판문점내 헬기 비행 권한 행사를 위한 월1회 판문점 헬기 비행 훈련(JSA-판문점 H-128), 서북도서 해상사격훈련 참관 등 작전활동을 하고 있습니다.

특별조사활동
Special Investigation

유엔군사령관이 주관하는 대사급 회의는 군사정전위원회 위원들과 회원국 대사를 대상으로 현 안보상황, 유엔군측 주요추진업무 등을 공유하는 회의입니다.

또한 군사정전위원회 수석대표는 유엔사 회원국 연락단과 한국군 연락단으로 구성된 군사정전위원회 자문단 회의를 실시하여 특별조사결과와 GP 점검결과 등 군사정전위원회 주요추진업무 및 협조사항을 공유하고 있습니다. 그리고 유엔군사령부 참모장과 중립국감독위원회 및 군사정전위원회 비서처와 함께 집행위원회를 운영하여 중립국감독위원회의 주요 추진업무 및 협조사항을 토의하고 중립국감독위원회의 업무수행 여건을 보장하고 있습니다.

회원국 대사급 회의
UNC Ambassadors Roundtable

샤프(Walter L. Sharp) 전 유엔군사령관은 북한의 도발, 공격 및 1953년 정전협정 위반에 대응하기 위해 군사정전위원회의 임무는 매우 중요하다고 강조하면서 군사정

commended the MAC's activities such as special investigations and follow-on reports to the UN Security Council on the sinking of the Cheonan and the shelling of Yeonpyeong-Do.

In 2014, GEN Curtis Scaparrotti, then the commander of the UNC, announced to revitalize the UNC. The UNC Revitalization was a policy to split multi-hatted positions to have them more dedicated to individual commands includong the UNC and expand the role of UNC member states. This effort continues to date. It was pursued to improve the functional effectiveness and roles of the command as many of the UNC staff members at the time were also hatted as CFC and/or USFK staffs, which caused confusion as to under which command they were conducting a task.

In addition, the necessity for splitting hats became more apparent with the relocation of the UNC and USFK to Camp Humphreys, physically separating the two commands from the CFC, as it would make it more challenging for the staffs to work for multiple commands that are geographically separated.

Above all, it was driven with an anticipation that boosting the UNC member states' role and function would greatly contribute to achieving a lasting peace on the Korean Peninsula (Interview with Deputy Commander, UNC).

Another former UNC Commander, GEN Vincent K. Brooks, submitted an annual report of the UNC activities to the UN Security Council (Report enclosed).

To reinforce the structure and capabilities of the UNC, a Canadian Lieutenant General was recently appointed to the position of the

전위원회는 천안함 피격 및 연평도 포격도발과 관련해 특별조사를 실시하고 유엔 안전보장이사회에 조사결과를 보고하는 등 임무를 수행했다고 높이 평가하였습니다.

2014년 커티스 스캐퍼로티 전 유엔군사령관은 유엔사 재활성화(UNC Revitalization) 프로그램을 공식적으로 표명하였습니다. 이는 유엔사 참모보직을 겸직에서 독립시키고 유엔사 회원국의 역할을 확대하는 정책이었으며, 현재까지도 진행하고 있습니다.

이러한 유엔사 재활성화의 배경은 유엔사의 역할과 기능 수행의 효율성을 제고하기 위해서였습니다. 당시 많은 참모들이 유엔군사령부, 연합군사령부, 주한미군사령부의 참모직책을 겸임하고 있었습니다.

따라서 임무가 부여되면 어떠한 신분의 입장에서 임무를 수행해야 하는지가 충분히 이해하지 못하는 경우가 있었고 또한 유엔군사령부와 주한미군사령부의 평택 이전으로 연합사령부와는 멀리 떨어져 참모가 겸직으로 임무수행하는 것이 쉽지 않게 되었습니다.

무엇보다도 유엔군사령부 회원국들의 역할과 기능을 확대하는 것이 한반도의 항구적 평화달성에 크게 기여할 것이라는 기대가 있었습니다.(유엔군부사령관 인터뷰)

브룩스(Vincent K. Brooks) 전 유엔군사령관은 유엔 안보리 제출용 유엔군사령부 연례보고서를 작성하여 제출

Deputy Commander and a U.S. Major General as its Chief of Staff, further revitalizing the command.

Also, cooperation with the UNC has greatly increased for the implementation of the CMA agreed between the North and the South on September 19, 2018 based on the spirit of the Armistice Agreement.

하였습니다(관련내용은 첨부하였음).

최근 유엔군 편성과 기능보강의 일환으로 캐나다 출신의 중장이 부사령관으로 보직되어 임무 수행중에 있고, 미국 출신 소장이 참모장으로 보직되는 등 유엔군사령부 기능이 크게 활성화 되었고, 정전협정의 정신을 바탕으로 2018년 9월 19일 합의된 남북한의 군사분야합의서 이행에 있어 유엔군사령부의 협조분야가 크게 증가하고 있습니다.

2. UN Resolution on the Korean War

The Requestment of the Cessation of Hostilities in Korea

June 25, 1950

The Security Council,

Recalling the finding of the General Assembly in its resolution of 21 October 1949 that the Government of the Republic of Korea is a lawfully established government "having effective control and jurisdiction over that part of Korea where the United Nations Temporary Commission on Korea was able to observe and consult and in which the great majority of the people of Korea reside; and that this Government is based on elections which were a valid expression of the free will of the electorate of that part of Korea and which were observed by the Temporary Commission; and that this is the only such Government in Korea";

Mindful of the concern expressed by the General Assembly in its resolution of 12 December 1948 and 21 October 1949 of the consequences which might follow unless Members States refrained from acts derogatory to the results sought to be achieved by the United nations in bringing about the complete independence and unity of Korea; and the concern expressed that the situation described by the United Nations Commission on Korea in its report menaces the safety military conflict there;

Nothing with grave concern the armed attack upon the Republic of Korea by forces from North Korea,

2. 한국전쟁과 관련된 유엔 안전보장이사회 결의

전쟁행위 중지에 관한 유엔 결의안 82호

1950년 6월 25일

안전보장이사회는,

대한민국 정부가 "국제연합 한국임시위원단이 감시 및 협의할 수 있었고 대한민국의 대다수가 거주하고 있는 한국지역에 관한 효과적인 통치력과 사법권을 갖고 있는" 수립된 정부라는 것과 "또한 이 정부가 한국의 이 지역의 유권자의 자유의사의 정당한 표현이며 임시위원단이 감시한 선거에 기초를 두고 있다는 것, 또 이 정부가 한국에 있어서 유일한 정부라는 것" 임을 결의한 1949년 10월 21일 총회결의를 상기하고, 한국의 안전한 독립과 통일을 가져옴에 있어서 국제연합이 이룩하고자 추구하는 성과에 유해한 재행위를 회원국이 삼가지 않는 한 초래될지도 모르는 결과에 관하여, 총회가 1948년 12월 12일 및 1949년 10월 21일 결의에서 표명한 우려와 국제연합 한국위원단이 그 보고서에서 기술한 사태가 대한민국 안전과 한국 민족의 복지 위협 및 한국에서의 군사적 충돌을 야기할지도 모른다고 표명한 우려를 유의하며, 북한군의 대한민국에 대한 무력공격을 중대한 관심을 갖고 주목하고, 이 행동이 평화를 파괴하는 요인이 된다고 단정하여,

Determines that this action constitutes a breach of the peace,

1. Calls for the immediate cessation of hostilities; and calls upon the authorities of North Korea to withdraw forthwith their armed forces to the thirty-eighth parallel;
2. Requests the United Nations Commission on Korea

   a) To communicate its fully considered recommendations on the situation with the least possible delay,

   b) To observe the withdrawal of the North Korea forces to the thirty-eight parallel, and

   c) To keep the Security Council informed on the execution of this resolution;
3. Calls upon all Members to render every assistance to the United Nations in the execution of this resolution and to refrain from giving assistance to the North Korean authorities.

The Military Assistance to Republic of Korea

June 27, 1950

The Security Council,

Having determined that the armed attack upon the Republic of Korea by forces from North Korea constitutes a breach of the peace,

Having called for an immediate cessation of hostilities and Having called upon the authorities of North Korea to withdraw forthwith their armed forces to the thirty-eighth parallel, and Having noted from the report of the united Nations Commission for Korea that the authorities in North Korea have neither ceased hostilities nor withdrawn their armed forces to the thirty-eighth parallel, and that urgent military

1. 전쟁행위의 즉각적인 중지를 요구하고, 북한 당국은 즉시 그 군대를 북위 38도선까지 철수시킬 것을 요구한다.
2. 국제연합 한국위원단에 대하여
   1) 이 사태에 관하여 충분히 검토된 재건의를 가능한 한 지체 없이 통지할 것
   2) 북위 38도선까지의 북한군의 철수를 감시할 것

   3) 본 협의의 이행에 관하여 계속 안전보장이사회에 통지할 것을 요청한다.
3. 모든 회원국에 대하여 본 결의의 이행에 있어서 국제연합의 모든 지원을 제공할 것과 북한당국에 대한 지원제공을 삼갈 것을 요청한다.

대한민국에 대한 군사원조에 관한 유엔 결의안 83호

1950년 6월 27일

안전보장이사회는,

북한군의 대한민국에 대한 무력공격이 평화를 파괴한 요인이 된다고 단정하고, 전쟁행위의 즉각적인 중지를 요구하고 북한 당국이 그 군대를 즉시 북위 38도선까지 철수시키지도 않았다는 것과 국제적인 평화와 안전을 회복시키는 데 긴급한 군사적 조치가 요청된다는 국제연합 한국위원단의 권고서를 주목하고, 평화와 안전을 확보하기 위하여 즉각적이고 효과적인 조치를 국제연합에 요청한 대한민국의 호소를 주목하여, 국제연합 회원

measures are required to restore international peace and security, and Having noted the appeal from the Republic of Korea to the United Nation for immediate and effective steps to secure peace and security,

Recommends that the Members of the United Nations furnish such assistance to the Republic of Korea as may be necessary to repel the armed attack and to restore international peace and security in the area.

The Creation and Operation of the Unified Command

July 7, 1950

The Security Council,

Having determined that the armed attack upon the Republic of Korea by forces from North Korea constitutes a breach of the peace,

Having recommended that Members of the United Nations furnish such assistance to the Republic of Korea as may be necessary to repel the armed attack and to restore international peace and security in the area,

1. Welcomes the prompt and vigorous support which governments and peoples of the United Nations have given to its Resolutions of 25 and 27 June 1950 to assist the Republic of Korea in defending itself against armed attack and thus to restore international peace and security in the area;
2. Notes that Members of the United Nations have transmitted to the United Nations offers of assistance for the Republic of Korea;
3. Recommends that all members providing military forces and other assistance pursuant to the aforesaid Security council resolutions

국들이 대한민국에 대하여, 이 지역에 무력공격을 격퇴하고 국제평화와 안전을 회복하기 위하여 필요한 지원을 제공할 것을 권고한다.

국제연합통합군사령부 설치에 관한 유엔 결의안 84호

1950년 7월 7일

안전보장이사회는,

북한군의 대한민국에 대한 무력공격이 평화를 파괴하는 요인이 되고 있다고 단정하고, 국제연합 회원국이 대한민국에 대하여, 이 지역에서 무력공격을 격퇴하고 국제평화와 안전을 회복하기 위하여 필요한 지원을 제공할 것을 권고하였으므로,

1. 국제연합의 각 정부와 국민이 무력공격에 대한 대한민국의 자위노력을 지원하고 이에 따라 이 지역에 국제평화와 안전을 회복하기 위한 1950년 6월 25일 및 27일 결의에 대하여 표명한 신속하고 강력한 지지를 환영한다.
2. 국제연합 회원국들이 대한민국에 대한 국제연합의 지원결의에 호응하였음을 주목한다.
3. 전술한 안전보장이사회의 재결의에 의거하여 병력 기타 지원을 제공하는 모든 회원국은 이러한 병력 기타

make such forces and other assistance available to a unified command under the United Sates;

4. Requests the United States to designate the commander of such forces;
5. Authorizes the unified command at its discretion to use the United Nations flag in the course of operations against North Korean forces concurrently with the flags of the various nations participating;
6. Requests the United States to provide the Security Council with reports as appropriate on the course of action taken under the unified command.

Exchange of Public Letters Concerning Transfer of Operation Command

① The Korean President Syngman Rhee to the American Embassy(July 15, 1950)

Dear General macArthur:

In view of the common military effort of the United Nations on behalf of the Republic of Korea in which all military forces, land, sea and air, of all the United Nations fighting in or near Korea have been placed under your operational command, and in which you have been designated Supreme Commander United Nations Forces, I am happy to assign to you have been authority over all land, sea and air forces of the Republic of Korea during the period of the continuation of the present state of hostilities, such command to be exercised either by you

지원을 미국 주도하의 통합군사령부가 이용할 수 있도록 할 것을 권고한다.

4. 미국에 대하여 이러한 군대의 사령관을 임명할 것을 위임한다.
5. 통합군사령부에 대하여는 북한군에 대한 작전중 참전 각국 국기와 함께 국제연합기를 임의대로 병용할 권한을 부여한다.
6. 미국에 대하여 통합군사령부 지휘하에 취해지는 활동 과정에 관하여 적절한 시기에 안전보장이사회 권고서를 제출할 것을 요청한다.

한국 육 · 해 · 공군 작전지휘권 이양에 관한 이승만 대통령과 맥아더 유엔군 총사령관간의 교환 공한

1950년 7월 14일

① 이 대통령이 맥아더 장군에게 보낸 공한(1950년 7월 15일)

맥아더 장군 귀하

대한민국을 위한 국제연합의 공동 군사 노력에 있어 한국내 또는 한국 근해에서 작전중인 국제연합의 모든 부대는 귀하의 통솔하에 있으며 또한 귀하는 그 최고사령관으로 임명되어 있음에 감하여, 본인은 현 작전상태가 계속되는 동안 일절의 지휘권을 이양하게 된 것을 기쁘게 여기는 바이오며, 여사한 지휘권은 귀하 자신 또는 한국 근해에서 행사하도록 위임한 기타 사령관이 행사하여야 할 것입니다.

personally or by such military commander or commanders to whom you may delegate the exercise of this authority within Korea or in adjacent seas.

The Korea Army will be proud to serve under your command, and the Korea people and Government will be equally proud and encouraged to have the overall direction of our combined combat effort in the hands of so famous and distinguished a soldier who also in his person possesses the delegated military authority of all the United Nations who have jointed together to resist this infamous communist assault on the independence and integrity of our beloved land.

With continued highest and warmest feelings of personal regard,

Sincerely your,
Syngman Rhee

② Reply of General MacArthur to President Rhee through American Ambassador John J. Muccio in Korea

July 16, 1950, Taegu, Korea

Dear Mr. President:

I am happy to transmit to you the following message from General of the Army Douglas MacArthur with reference to your letter of July in which you designated to his operational command authority over the land, sea and air forces of the Republic of Korea during the present hostilities.

"Please express to President Rhee my thanks and deepest appreciation for the action taken in his letter of 15 July. It cannot fall to

한국군은 귀하의 휘하에서 복무하는 것을 영광으로 생각할 것이며 또한 한국 국민과 정부도 고명하고 훌륭한 군인으로서, 우리들의 사랑하는 국토의 독립과 보전에 대한 비열한 공산침략에 대항하기 위하여 힘을 합친 국제연합의 모든 군사권을 받고 있는 귀하의 전체적 지휘를 받게 된 것을 영광으로 생각하며 또한 격려되는 바입니다.

귀하에게 심후하고도 따뜻한 개인적인 경의를 표하나이다.

1950년 7월 14일
이승만

② 맥아더 장군의 회한(주한 미국대사를 통해 전달 : 1950년 7월 18일)

1950년 7월 16일

대통령 각하

현 적대상태가 계속되는 동안 대한민국 육·해·공군의 작전지휘권을 위임한 7월 14일부 귀하의 서신에 관한 맥아더 원수의 다음과 같은 회신을 전달함은 본관은 영광으로 생각합니다.

7월 15일자 공한에 의하여 이 대통령이 취하신 조치에 대하여 본관의 충심으로부터의 감사와 심심한 신의를 그에게 표하여 주심을 바라나이다. 한국내에서 작전중

increase the coordinated power of the United Nations Forces operating in Korea. I am proud indeed to have the gallant Forces of the Republic of Korea under my command. Please tell President I am grateful for his generous references to me personally and how sincerely I reciprocate his sentiments of regard. Tell him also not to lose heart, that the way may be long and hard, but the ultimate result cannot fail to be victory.

MacArthur."

With assurances of my ever continued highest regards, I remain

Sincerely Your,

John J. muccio

The Honorable Syngman Rhee,

President of the Republic of Korea, Taegu

Setting Up a Cease-Fire Group of 3 Persons

December 14, 1950

The General Assembly,

Viewing with grave concern the situation in the Far East,

Anxious that immediate steps should be taken to prevent the conflict in Korea spreading to other areas and to put an end to the fighting in Korea itself, and that further steps should then he taken for a peaceful settlement of existing issues in accordance with the Purposes and Principles of the United Nations.

Requests the President of the General Assembly to constitute a group of three persons, including himself, to determine the basis on which a

인 국제연합군의 통솔력은 반드시 증강될 것입니다. 용감무쌍한 대한민국군을 본관 지휘하에 두게 된 것을 영광으로 생각하나이다. 이 대통령의 본관에 대한 과도한 개인적 찬사에 대한 사의와 그에 대하여 본관이 또한 가지고 있는 존경의 뜻도 아울러 전달하여 주시기 바라나이다. 우리들의 장래가 고난하고 요원할지도 모르겠으나 종국적 결과는 반드시 승리할 것이므로 실망하지 마시도록 그에게 전언하여 주시기 바랍니다.

맥아더"

본인의 변함없는 존경과 함께

죤 · 무치오

대한민국 대통령

이승만 각하

정전3인위원회 설치에 관한 결의*

1950년 12월 14일

총회는,

극동사태를 중대한 관심을 가지고 고찰하며,

한국분쟁이 다른 지역에까지 확대함을 방지하고 한국전쟁 자체를 종식시키기 위한 대책을 즉시 취할 것과 '유엔'의 목적과 원칙에 입각하여 잔존문제를 평화적으로 해결할 차후 대책을 취할 것을 열망하며, 총회 의장에 대하여 의장 자신을 포함한 3인위원단을 구성하여 한국의 만족할 만한 정전기초를 결정하고 가급적 신속히 총회에 건의 제출할 것을 요청한다.**

*한국정전에 관한 유엔 결의안

**이 결의문은 제 325차 총회에서 52대 5, 기권 1표로 가결되었다. 정전위원회는 나꼬돌라 · 엔테잠 의장, A.L · B 피어슨 캐나다 외상 및 인도 수석대표 B · N · 라우 경의 3명으로 조직되었다.

satisfactory cease-fire in Korea can be arranged and to make recommendations to the General Assembly as soon as possible.

Joint Policy Declaration Concerning the Korean Armistice

July 27, 1953
Issued at Washington

We, the United Nations Members whose military forces are participating in the Korean action, support the decision of the Commander-in-Chief of the United Nations Command to conclude an armistice agreement. We hereby affirm our determination fully and faithfully to carry our the terms of that armistice. We expect that the other parties to the agreement will likewise scrupulously observe its terms.

The task ahead is not an easy one. We will support the efforts of the United Nations to bring about an equitable settlement in Korean based on the principles which have long been established by the United Nations, and which call for a united, independent and democratic Korea. We will support the United Nations in its efforts to assist the people of Korea in repairing the ravages of war.

We declare again our faith in the principles and purposes of the United Nations, our consciousness of our continuing responsibilities in Korea, and our determination in good faith to a renewal of the armed attack, challenging again the principles of the United Nations, we should again be united and prompt to resist. The consequences of such a breach of the armistice would be so grave that, in all probability, it would not be possible to confine hostilities within the frontiers of

한국휴전에 관한 참전 16개국 공동정책 선언

1953년 7월 27일
미국 워싱턴에서 발표

한국전쟁에 군대를 파견하고 있는 우리 국제연합 회원국은 휴전협정을 체결하기 위한 유엔군 총사령관의 결정을 지지한다.

우리는 동 휴전협정의 제 조항을 전폭적이며 성실하게 이행하려는 우리의 결의를 확인한다.

우리는 협정의 상대측도 우리와 같이 제 조항을 성실히 수행할 것을 기대한다.

우리는 전화(戰禍)를 복구함에 있어서 한국민을 원조하려는 국제연합의 제 노력을 지지한다.

우리는 국제연합의 제 목적과 원칙에 대한 신임과 한국에 있어서의 우리들의 변함없는 책임에 대한 인식 및 한국문제 해결을 성실하게 추구하려는 우리들의 결의를 재천명한다.

만약 국제연합 제 원칙에 반한 무력공격이 재발한 경우, 우리는 세계평화를 위하여 다시 단결하여 즉각적으로 이에 대항할 것임을 확인한다.

이와 같은 휴전협정 위반의 결과는 필경 전쟁을 한국 전선내에서 제한할 수 없을 만큼 중대한 것이 되게 할 것이다.

끝으로 우리는 이 휴전협정이 아시아의 기타 여사한

Korea. Finally, we are of the opinion that the armistice must not result in jeopardizing the restoration or the safeguarding of peace in any other part of Asia.

Australia : Percy C. Spende
The Nertherlands : J. H. van Roijen
Belgium : Silvercruys
New Zealand : L. K. Munro
Canada : H. H. Wrong
The Philippines : Melquaides J. Gamboa
Colombia : Cipriano Restrepo J.
Thailand : P. Sarasin
Ethiopia : A. Tesemama
Turkey : Feidum C. Erkin
France : H. Bonnet
The Union of South Africa : G. P. Jooste
Greece : A. G. Politis
The United Kingdom : Roger makins
Luxembourg : Hugues le Gallais
United States of America : John Foster Dulles

Sixteen-Nation Declaration of Korea Made at Close of the Korean Phase of the Geneva Conference

June 15, 1954, Issued at Geneva

Pursuant to the resolution of August 28, 1953, of the United Nations General Assembly, and the Berlin communique of February 18, 1954, we as nations who contributed military forces to the United Nations Command in Korea, have been participating in the Geneva Conference for the purpose of establishing a united and independent Korea by peaceful means.

We have made a number of proposals and suggestions in accord with the past efforts of the United Nations to bring about the unification, independence, and freedom of Korea; and within the framework of the following two principles which we believe to be

지역에 있어서의 평화의 회복과 보장을 위태롭게 하는 결과가 되어서는 안 된다고 생각한다.

호주	에티오피아
네덜란드	터키
벨기에	프랑스
뉴질랜드	남아공
캐나다	그리스
필리핀	영국
콜롬비아	룩셈부르크
태국	미국

한국참전 16개국의 제네바 회담 공동성명

1954년 6월 15일 스위스 제네바에서 발표

1953년 8월 28일 UN총회의 결의와 1954년 2월 18일의 베를린 「코뮈니케」에 준거하여 주한 UN군에 군사력을 제공한 국가들로서, 우리는 평화적 방법에 의하여 통일된 독립 한국을 수립하기 위하여 제네바 회담에 참가하고 있다.

우리는 한국의 통일과 독립의 자유를 가져오기 위하여 UN이 과거에 해 온 노력에 입각하여 기본적으로 생각되는 다음과 같은 두 원칙의 범위내에서 많은 제공과 제의를 하였던 것이다.

fundamental.

1. The United Nations, under its Charter, is fully and rightfully empowered to take collective action to repel aggression, to restore peace and security, and to extend its good offices to seeking a peaceful settlement in Korea.
2. In order to establish a unified, independent and democratic Korea, genuinely free elections should held under UN supervision, for representatives in the national assembly, in which shall be in direct proportion to the indigenous population in Korea. We have earnestly and patiently searched for a basis of agreement which would enable us to proceed with Korean unification in accordance with these fundamental principles. The Communist delegations have rejected out every effort to obtain agreement. The principal issues between us, therefore, are clear. Firstly, we accept and assert the authority of the authority of the United Nations. The Communists repudiate and reject the authority and competence of the United Nations in Korea and have labelled the United Nations itself as the tool of aggression. Were we to accept this position of the Communists, it would mean the death of the principle of collective security and of the UN itself. Secondly, we would desire genuinely free elections. The Communists insist upon procedures which would make genuinely free elections impossible. It is clear that the Communists will not accept impartial and effective supervision of free elections. They have shown their intention to maintain Communist control over North Plainly, Korea. They have persisted in the same attitudes which have frustrated United Nations efforts to unify Korea since 1947.

We believe, therefore, that it is better to face the fact of our disagreement than to raise false hopes and mislead the peoples of

1. UN은 그 헌장에 의거하여 침략을 격퇴하기 위하여 집단행동을 취하고 평화와 안전을 회복하고 한국에서의 평화적 해결을 강구하기 위하여 간여할 전적이며 정당한 권한이 부여되어 있다.
2. 통일되고 독립된 민주주의 한국을 수립하기 위하여, UN의 감독하에 진정한 자유선거가 이루어져야 하며 국회의원 선출 방식은 한국 인구의 직접 비례에 의한 대표방식이어야 한다. 우리는 이러한 기본원칙에 입각하여 한국의 통일을 유지시키기 위하여 꾸준히 열성적으로 노력하여 왔으나, 공산대표단들은 합의를 얻으려는 우리의 모든 노력을 거절하여 왔다. 그러므로 우리들 사이에 개재하는 원칙적인 문제는 명백하다. 첫째로, 우리는 UN의 권위를 수락하고 또 주장한다. 공산측은 UN의 이 권위와 함께 한국에 있어서의 UN의 권력을 부인하고 거절하였으며, UN 자체를 침략의 도구라고 낙인찍었다. 만약 우리가 공산측의 이러한 입장을 수락한다면 이것은 집단안전의 원칙과 UN 자체의 멸망을 의미하는 것이다.
   둘째로, 우리는 진정한 자유선거를 원하고 있으나 공산측은 진정한 자유선거를 불가능하게 할 절차를 고집하고 있다. 공산측이 자유선거의 공평하고도 효과적인 감시를 수락하지 않을 것이 명백하며, 그들은 북한을 계속적으로 지배하려는 그들의 의도를 명시하였다. 그들은 1947년 이래로 한국을 통일시키려는 UN의 노력을 좌절시켜 온 태도를 그대로 고집하고 있다.

그러므로 우리는 합의가 없는 곳에 합의가 있는 듯이 그릇된 희망을 가지게 하여 세계의 인민들을 오도하

the world into believing that there is agreement where there is none.

In the circumstances, we have been compelled reluctantly and regretfully to conclude that so long as the Communist delegations reject the two fundamental principles which we consider indispensable, further consideration and examination of the Korean question by the conference would serve no useful purpose. We reaffirm our continued support for the objectives of the United Nations in Korea.

In accordance with the resolution of the General Assembly of the United Nations of August 28, 1953, the member states parties to this declaration will inform the United Nations concerning the proceedings at this conference.

기보다는 오히려 우리의 의견이 수립되어 있다는 사실을 직시하게 하는 것이 더 낫다고 믿는다. 이러한 태도하에서 우리는 우리가 불가결한 것으로 생각하는 두 개의 기본원칙을 공산대표들이 거절하는 한 이 회담에서 한국문제를 이 이상 더 심의 고려하는 것은 소용없는 일이라는 결론을 부득이 내리지 않을 수 없는 것이다.

우리는 한국에 있어서의 UN의 목적을 계속 지지할 것을 재확인하는 바이다. 1953년 8월 28일 UN총회의 결의에 의거하여 이 선언서에 서명한 각 회원국은 이 회담의 의사에 관하여 UN에 통고할 것이다.

3. ARMISTICE AGREEMENT

AGREEMENT BETWEEN THE COMMANDER-IN-CHIEF, UNITED NATIONS COMMAND, ON THE ONE HAND, AND THE SUPREME COMMANDER OF THE KOREAN PEOPLE'S ARMY AND THE COMMANDER OF THE CHINESE PEOPLE' S VOLUNTEERS, ON THE OTHER HAND, CONCERNING A MILITARY ARMISTICE IN KOREA

PREAMBLE

The undersigned, the Commander-in-Chief, United Nations Command, on the one hand, and the Supreme Commander of the Korean People's Army and the Commander of the Chinese People's Volunteers, on the other hand, in the interest of stopping the Korean conflict, with its great toll of suffering and bloodshed on both sides, and with the objective of establishing an armistice which will insure a complete cessation of hostilities and of all acts of armed force in Korea until a final peaceful settlement is achieved, do individually, collectively, and mutually agree to accept and to be bound and governed by the conditions and terms of armistice set forth in the following Articles and Paragraphs, which said conditions and terms are intended to be purely military in character and to pertain solely to the belligerents in Korea.

3. 정전협정 전문

국제연합군 총사령관을 일방으로 하고 조선인민군 최고사령관 및 중국인민지원군 사령원을 다른 일방으로 하는 한국 군사 정전에 관한 협정

- 국제연합군 총사령관 미국 육군 대장 : 마크 더블유 클라크(Mark W. Clark)
- 조선인민군 최고사령관 조선민주주의인민공화국 원수 : 김일성
- 중국인민지원군 사령관 : 팽더화이(彭德懷)

서언

국제연합군 총사령관을 일방으로 하고 조선인민군 최고사령관 및 중국인민지원군 사령원을 다른 일방으로 하는 하기의 서명자들은 쌍방에 막대한 고통과 유혈을 초래한 한국 충돌을 정지시키기 위하여 최후적인 평화적 해결이 달성될 때까지 한국에서의 적대행위와 일체 무장행동의 완전한 정지를 보장하는 정전을 확립할 목적으로 하기 조항에 기재된 정전 조건과 규정을 접수하며 또 그 제약과 통제를 받는데 각자 공동 호상 동의한다.

이 조건과 규정들의 의도는 순전히 군사적 성질에 속하는 것이며 이는 오직 한국에서의 교전 쌍방에만 적용한다.

- 한국충돌 : 1950년 6월 25일 새벽에 북한 공산군이 남북 군사분계선이던 38선 전역에 걸쳐 불법 남침함으로써 일어난 6.25한국전쟁입니다.

ARTICLE 1
MILITARY DEMARCATION LINE AND DEMILITARIZED ZONE

1. A Military Demarcation Line shall be fixed and both sides shall withdraw two(2) kilometers from this line so as to establish a Demilitarized Zone between the opposing forces. A Demilitarized Zone shall be established as a buffer zone to prevent the occurrence of incidents which might lead to a resumption of hostilities.

2. The Military Demarcation Line is located as indicated on the attached map(Map 1).

3. The Demilitarized Zone is defined by a northern and a southern boundary as indicated on the attached map(Map 1).

4. The Military Demarcation Line shall be plainly marked as directed by the Military Armistice Commission hereinafter established. The Commanders of the opposing sides shall have suitable markers erected along he boundary between the Demilitarized Zone and their respective areas. The Military Armistice Commission shall supervise the erection of all markers placed along the Military Demarcation Line and along the boundaries of the Demilitarized Zone.

5. The waters of the Han River Estuary shall be open to civil shipping of both sides wherever one bank is controlled by one side and the other bank is controlled by the other side. The Military Armistice Commission shall prescribe rules for the shipping in that part of the Han River Estuary indicated on the attached map(Map 2). Civil

제1조
군사분계선과 비무장지대

1. 한 개의 군사분계선을 확정하고 쌍방이 이 선으로부터 각기 이(2)킬로미터씩 후퇴함으로써 적대군대간에 한 개의 비무장지대를 설정한다. 한 개의 비무장지대를 설정하여 이를 완충지대로 함으로써 적대행위의 재발을 초래할 수 있는 사건의 발생을 방지한다.

2. 군사분계선의 위치는 첨부한 지도에 표시한 바와 같다.(지도1－본서 pp.192～200 참조)

3. 비무장지대는 첨부한 지도에 표시한 북경계선 및 남경계선으로써 이를 확정한다.(지도1－본서 pp.192～200 참조)

4. 군사분계선을 하기와 같이 설립한 군사정전위원회의 지시에 따라 이를 명백히 표식한다. 적대쌍방 사령관들은 비무장지대와 각자의 지역간의 경계선에 따라 적당한 표식물을 세운다. 군사정전위원회는 군사분계선과 비무장지대의 량 경계선에 따라 설치한 일체 표식물의 건립을 감독한다.

5. 한강 하구의 수역으로서 그 한쪽 강안이 일방의 통제하에 있고 그 다른 한쪽 강안이 다른 일방의 통제하에 있는 곳은 쌍방의 민선박의 항행에 이를 개방한다. 첨부한 지도(첨부한 지도 제2도－본서 p.201을 보라.)에 표시한 부분의 한강 하구의 항행 규칙은 군사정전위

• 군사분계선 : 남북의 군사 경계선을 말하며 이것이 이른바 휴전선입니다. 그 길이는 모두 155마일(약 250km)로, 서해안의 강화(江華) 북방에서 동해안의 간성(杆城) 북방까지입니다.

• 비무장지대 : 군사 분계선 남북 양쪽 2km의 대상지역을 비무장지대로 설정하여 이를 완충지대화 함으로써 적대행위의 발생을 방지합니다.

• 경계선의 표식물은 임진강 하구 장단반도에 제0001호에서 시작하여 동해안의 마지막 표식물 1,292호까지임. 그중 696개는 유엔군측이 나머지 596개는 공산군측이 관리하기로 하였습니다.

• 한강 하구의 수역 : 한강에 설정한 비무장지대(DMZ)로서 만우리에서 강화군 서도면 볼음도까지 약 67km 구간을 중립수역으로 정했습니다. 분쟁 가능성이 높은 민감한 지역으로 유엔군사령부 군사정전위원회가 관리·통제하고 있다. 1953년 당시 강원도에서 시작된 군사분계선은 경기도 파주시 탄현면 만우리까지만 설정되어 한강 하구에는 별도의 완충지역이 필요했습니다.

shipping of each side shall have unrestricted access to the land under the military control of that side.

6. Neither side shall execute any hostile act within, from, or against the Demilitarized Zone.

7. No person, military or civilian, shall be permitted to cross the Military Demarcation Line unless specifically authorized to do so by the Military Armistice Commission.

8. No person, military or civilian, in the Demilitarized Zone shall be permitted to enter the territory under the military control of either side unless specifically authorized to do so by the Commander into whose territory entry is sought.

9. No person, military or civilian, shall be permitted to enter the Demilitarized Zone except persons concerned with the conduct of civil administration and relief and persons specifically authorized to enter by he Military Armistice Commission.

10. Civil administration and relief in that part of the Demilitarized Zone which is south of the Military Demarcation Line shall be the responsibility of the Commander-in-Chief, United Nations Command; and civil administration and relief in that part of the Demilitarized Zone which is north of the Military Demarcation Line shall be the joint responsibility of the Supreme Commander of the Korean People's Army and the Commander of the Chinese People's

원회가 이를 규정한다. 각방 민선박이 항행함에 있어서 자기측의 군사통제하에 있는 육지에 배를 대는 것은 제한받지 않는다.

6. 쌍방은 두 비무장지대내에서 또는 비무장지대로부터 또는 비무장지대에 향하여 어떠한 적대행위도 감행하지 못한다.

7. 군사정전위원회의 특정한 허가없이는 어떠한 군인이나 사민이나 군사분계선을 통과함을 허가하지 않는다.

8. 비무장지대내의 어떠한 군인이나 사민이나 그가 들어갈려고 요구하는 지역의 사령관의 특정한 허가없이는 어느 일방의 군사통제하에 있는 지역에도 들어감을 허가하지 않는다.

9. 민사행정 및 구제사업의 집행에 관계되는 인원과 군사정전위원회의 특정한 허가를 얻고 들어가는 인원을 제외하고는 어떠한 군인이나 사민이나 비무장지대에 들어감을 허가하지 않는다.

10. 비무장지대내의 군사분계선 이남의 부분에 있어서의 민사행정 및 구제사업은 국제연합군 총사령관이 책임진다. 비무장지대내의 군사분계선 이북의 부분에 있어서의 민사행정 및 구제사업은 조선인민군 최고사령관과 중국인민지원군 사령원이 공동으로 책임진다. 민사행정 및 구제사업을 집행하기 위하여 비무장지대에 들어갈 것을 허가받는 군인 또는 사민의 인원수는

- 군사정전위원회는 정전협정 이행을 감독하며 정전협정 위반사건 발생시 이를 협의하여 처리하는 쌍방 5명의 대표로 구성된 위원회입니다. 군사정전위원회는 정전협정이행 감독을 위해 구성되었으며 정전협정 위반사실이 있을 때마다 위원회를 소집하여 협의 처리함으로써 남북한간의 전쟁을 억제하고 한반도의 정세 안정에 기여해 왔습니다.
- 사민 : 일반인

Volunteers. The number of persons, military or civilian, from each side who are permitted to enter the Demilitarized Zone for the conduct of civil administration and relief shall be as determined by the respective Commanders, but in no case shall the total number authorized by either side exceed one thousand(1,000) persons at any one time. The number of civil police and the arms to be carried by them shall be as prescribed by the Military Armistice Commission. Other personnel shall not carry arms unless specifically authorized to do so by the Military Armistice Commission.

11. Nothing contained in this Article shall be construed to prevent the complete freedom of movement to, from, and within the Demilitarized Zone by the Military Armistice Commission, its assistants, its Joint Observer Teams with their assistants, the Neutral Nations Supervisory Commission hereinafter established, its assistants, its Neutral Nations Inspection Teams with their assistants, and of any other persons, materials, and equipment specifically authorized to enter the Demilitarized Zone by the Military Armistice Commission. Convenience of movement shall be permitted through the territory under the military control of either side over any route necessary to move between points within the Demilitarized Zone where such points are not connected by roads lying completely within the Demilitarized Zone.

각방 사령관이 각각 이를 결정한다. 단 어느 일방이 허가한 인원의 총수는 언제나 천(1,000)명을 초과하지 못한다. 민사행정 경찰의 인원수 및 그가 휴대하는 무기는 군사정전위원회가 이를 규정한다. 기타 인원은 군사정전위원회의 특정한 허가 없이는 무기를 휴대하지 못한다.

11. 본 조의 어떠한 규정이든지 두 군사정전위원회 그의 보조인원 그의 공동감시소조 및 소조의 보조인원 그리고 하기와 같이 설립한 중립국감독위원회 그의 보조인원 그의 중립국시찰소조 및 소조의 보조인원과 군사정전위원회로부터 비무장지대로 들어갈 것을 특히 허가받은 기타의 모든 인원, 물자 및 장비의 비무장지대 출입과 비무장지대내에서의 이동의 완전한 자유를 방해하는 것으로 해석하여서는 안 된다. 비무장지대내의 두 지점이 비무장지대내에 전부 들어 있는 도로로써 연결되지 않는 경우에 이 두 지점간 반드시 경과하여야 할 통로를 왕래하기 위하여 어느 일방의 군사통제하에 있는 지역을 통과하는 이동의 편리를 허여한다.

• 공동감시소조 : 정전위원회는 DMZ와 HRE 지역의 적대행위 위반여부 감시를 위해 공동감시소조를 두고 각 감시소조는 4명 내지 6명의 영관장교로 구성하되 쌍방이 각기 반수씩 임명하도록 되었으며, 처음에 10개의 감시소조를 두었으며, 1955년도에 5개 소조로 축소되었고, 1994년에 북한측의 군사정전위원회 철수로 기능이 거의 마비되었습니다.
1968년 유엔군사령관 지침에 의거 특별조사팀(SIT)이 조사를 하고 있습니다.

ARTICLE II
CONCRETE ARRANGEMENTS FOR CEASE-FIRE AND ARMISTICE

A. GENERAL

12. The Commanders of the opposing sides shall order and enforce a complete cessation of all hostilities in Korea by all armed forces under their control, including all units and personnel of the ground, naval, and air forces, effective twelve(12) hours after this Armistice Agreement is signed.(See Paragraph 63 hereof for effective date and hour of the remaining provisions of this Armistice Agreement.)

13. In order to insure the stability of the Military Armistice so as to facilitate the attainment of a peaceful settlement trough the holding by both sides of a political conference of a higher level, the Commander of the opposing sides shall:
a. Within seventy-two(72) hours after this Armistice Agreement becomes effective, withdraw all of their military forces, supplies, and equipment from the Demilitarized Zone except as otherwise provided herein. All demolitions, minefields, wire entanglements, and other hazards to the safe movement of personnel of the Military Armistice Commission or its Joint Observer Teams, known to exist within the Demilitarized Zone after the withdrawal of military forces there from, together with lanes known to be free of all such hazards, shall be reported to the Military Armistice Commission by the Commander of the side whose forces emplaced such hazards. Subsequently, additional safe lanes shall be cleared; and eventually,

제2조
정화 및 정전의 구체적 조치

가. 총칙

12. 적대쌍방 사령관들은 육해 공군의 모든 부대와 인원을 포함한 그들의 통제하에 있는 모든 무장역량이 한국에 있어서의 일체 적대행위를 완전히 정지할 것을 명령하고 또 이를 보장한다.
본 항의 적대행위의 완전정지는 본 정전협정이 조인된 지 십이(12) 시간 후부터 효력을 발생한다. (본 정전협정의 기타 각항의 규정이 효력을 발생하는 날짜와 시간에 대하여서는 본 정전협정 제63항을 보라.)

13. 군사정전의 확고성을 보장함으로써 쌍방의 한급 높은 정치회의를 진행하여 평화적 해결을 달성하는 것을 이룹게 하기 위하여 적대쌍방 사령관들은

ㄱ) 본 정전협정 중에 따로 규정한 것은 제외하고 본 정전협정이 효력을 발생한 후 칠십이(72) 시간내에 그들의 일체 군사역량, 보급 및 장비를 비무장지대로부터 철거한다. 군사역량을 비무장지대로부터 철거한 후 비무장지대내에 존재한다고 알려져 있는 모든 폭파물, 지뢰원, 철조망 및 기타 군사정전위원회 또는 그의 공동감시소조인원의 통행안전에 위험이 미치는 위험물들은 이러한 위험물이 없다고 알려져 있는 모든 통로와 함께 이러한 위험물을 설치한 군대의 사령관이 반드시 군사정전위원회에 이를 보고한다.
그 다음에 더 많은 통로를 청소하여 안전하게 만들어

• 정화 및 정전 : 교전 중 평화적 해결책을 찾기 위하여 일시적 합의에 따라 사격과 전투의 중지를 의미합니다.

• 정전협정이 조인된 시간은 1953년 7월 27일 10시이며 동일 22시부터 효력이 발생하였습니다.

• 협정 후 72시간 이내 비무장지대의 모든 장비를 철거하고 그리고 군사정전위원회의 감독하에 45일 이내 폭발물, 지뢰, 철조망 등 모든 위험물을 제거하게 하였습니다. 그러나 현재의 비무장지대는 지뢰와 철조망 그리고 GP로 무장되어 있습니다.

within forty-five(45) days after the termination of the seventy-two(72) hour period, all such hazards shall be removed from the Demilitarized Zone as directed by and under the supervision of the Military Armistice Commission. At the termination of the seventy-two(72) hour period, except for unarmed troops authorized a forty-five(45) day period to complete salvage operations under Military Armistice Commission supervision, such units of a police nature as may be specifically requested by the Military Armistice Commission and agreed to by the Commanders of the opposing sides, and personnel authorized under Paragraphs 10 and 11 hereof, no personnel of either side shall be permitted to enter the Demilitarized Zone.

b. Within ten(10) days after this Armistice Agreement becomes effective, withdraw all of their military forces, supplies, and equipment from the rear and the coastal islands and waters of Korea of the other side. If such military forces are not withdrawn within the stated time limit, and there is no mutually agreed and valid reason for the delay, the other side shall have the right to take any action which it deems necessary for the maintenance of security and order. The term "coastal islands", as used above, refers to those islands which, though occupied by one side at the time when this Armistice Agreement becomes effective, were controlled by the other side on 24 June 1950; provided, however, that all the islands lying to the north and west of the provincial boundary line between HWANGHJAE-DO and KYONGGI-DO shall be under the military control of the Supreme Commander of the Korean people's Army and the Commander of the Chinese People's Volunteers, except the island groups of PAENGYONG-DO(37° 58' N, 124° 40' E), TAECHONG-DO(37°50'N, 124°42'E), SOCHONG-DO(37°46'N,

결국에 가서는 칠십이(72) 시간의 기간이 끝난 후 사십오(45) 일내에 모든 이러한 위험물은 반드시 군사정전위원회의 지시에 따라 또 그 감독하에 비무장지대로부터 이를 제거한다. 칠십이(72) 시간의 기간이 끝난 후 군사정전위원회의 감독하에서 사십오(45) 일의 기간내에 제거작업을 완수할 권한을 가진 비무장부대와 군사정전위원회가 특히 요청하였으며 또 적대쌍방 사령관들이 동의한 경찰의 성질을 가진 부대 및 본 정전협정 제10항과 제11항에서 허가한 인원 이외에는 쌍방의 어떠한 인원이든지 비무장지대에 들어가는 것을 허가하지 않는다.

ㄴ) 본 정전협정이 효력을 발생한 후 십(10) 일 이내에 상대방의 한국에 있어서의 후방과 연해 섬들 및 해면으로부터 그들의 모든 군사역량, 보급물자 및 장비를 철거한다. 만일 철거를 연기할 쌍방이 동의한 이유없이 또 철거를 연기할 유효한 이유없이 기한이 넘어도 이러한 군사역량을 철거하지 않을 때에는 상대방은 치안을 유지하기 위하여 그가 필요하다고 인정하는 어떠한 행동이라도 취할 권리를 가진다. 상기한 '연해섬' 이라는 용어는 본 정전협정이 효력을 발생할 때에 비록 일방이 점령하고 있더라도 1950년 6월 24일에 상대방이 통제하고 있던 섬들을 말하는 것이다. 단 황해도와 경기도의 도계선 북쪽과 서쪽에 있는 모든 섬 중에서 백령도(북위 37도58분, 동경 124도40분), 대청도(북위 37도50분, 동경124도42분), 소청도(북위37도46분, 동경124도46분), 연평도(북위 37도38분, 동경125도40분) 및 우도(북위37도36분, 동경125도58분)의 도서군들을 국제연합군 총사령관의 군사통제하에 남

124°46'E), YONPYONG-DO(37°38'N, 125°40'E), and U-DO(37°36'N, 125°58'E), which shall remain under the military control of the Commander-in-Chief, United Nations Command. All the islands on the west coast of Korea lying south of the above-mentioned boundary line shall remain under the military control of the Commander-in-Chief, United Nations Command.(See Map 3.)

c. Cease the introduction into Korea of reinforcing military personnel; provided, however, that the rotation of units and personnel, the arrival in Korea of personnel on a temporary Duty basis, and the return to Korea of personnel after short period of leave or temporary duty outside of Korea shall be permitted within the scope prescribed below. "Rotation is defined as the replacement of units or personnel by other unites or personnel who are commencing a tour of duty in Korea. Rotation personnel shall be introduced into and evacuated from Korea only through the ports of entry enumerated in Paragraph 43 hereof. Rotation shall be conducted on a man-for-man basis; provided, however, that no more than thirty-five thousand(35,000) persons in the military service shall be admitted into Korea by either side in any calendar month under the rotation policy. No military personnel of either side shall be introduced into Korea if the introduction of such personnel will cause the aggregate of the military personnel of that side admitted into Korea since the effective date of this Armistice Agreement to exceed the cumulative total of the military personnel of that side who have departed from Korea since that date. Reports concerning arrivals in and departures from Korea of military personnel shall be made daily to the Military Armistice Commission and the Neutral Nations Supervisory Commission; such reports shall include places of arrival and departure and the number of

겨 두는 것을 제외한 기타 모든 섬들은 조선인민군 최고사령관과 중국인민지원군 사령원의 군사통제하에 둔다. 한국서 해안에 있어서 상기 경계선 이남에 있는 모든 섬들은 국제연합군 총사령관의 군사통제하에 남겨둔다.(지도3－본서 pp.202~203 참조)

ㄷ) 한국 경외로부터 증원하는 군사인원을 들어오는 것을 정지한다. 단 아래에 규정한 범위내의 부대와 인원의 윤환 임시임무를 담당한 인원의 한국에의 도착 및 한국경외에서 단기 휴가를 하였거나 혹은 임시임무를 담당하였던 인원의 한국에의 귀환은 이를 허가한다.

'윤환' (Rotation)의 정의는 부대 혹은 인원이 한국에서 복무를 개시하는 다른 부대 혹은 인원과 교체하는 것을 말하는 것이다.

윤환인원은 오직 본 정전협정 제43항에 열거한 출입항을 경유하여서만 한국에 들어오며 또 한국으로부터 내어갈 수 있다.

윤환은 일(1) 대 일(1)인의 교환 기초위에서 진행한다. 단, 어느 일방이든지 어느 일(1)력월 내에 윤환정책하에서 한국 경외로부터 삼만오천(35,000)명 이상의 군사인원을 들어오지는 못한다. 만일 일방의 군사인원을 들어오는 것이 해당측이 본 정전협정 효력 발생일로부터 한국으로 들어온 군사인원의 총수로 하여금 같은 날짜로부터 한국을 떠난 해당측의 군사인원의 누계 총수를 초과하게 할 때는 해당측의 어떠한 군사인원도 한국에 들어올 수 없다.

군사인원의 한국에의 도착 및 한국으로부터의 이거에 관하여 매일 군사정전위원회와 중립국감독위원회에 보고한다.

• 정전협정의 핵심중 하나로 남측과 북측의 새로이 군사력을 증강시키지 못하도록 감독하는 것입니다. 외부로부터 한반도로 군사인원과 작전비행기, 장갑차량, 무기 및 탄약이 들어오는 것을 통제하도록 하였습니다. 이를 위해 남측 5곳, 북측 5곳으로만 출입하도록 제한하고 중립국감독위원회가 중립국시찰소조를 활용하여 감독하도록 하였습니다.(남측 5곳; 인천, 대구, 부산, 강릉, 군산. 북측 5곳; 신의주, 청진, 흥남, 만포, 신안주)
그러나, 이 조항은 현재 지켜지지 못하고 있습니다. 북측은 정전 이후 재무장을 위하여 소련으로부터 무기와 장비를 들여오면서 북측에서 활동 중인 중립국감독위원회의 정상적인 감독과 보고 활동을 방해하였고, 남측에서 활동중이었던 북측의 중립국감독위원회는 간첩행위를 하였습니다. 이에 유엔사측은 1957년 6월 21일 군사정전위원회 제75차 본회의에서 북측의 정전협정 위반행위를 지적하고 북측이 한국으로의 무기 반입을 금지한 정전협정 제2조 13항의 준수의사를 표시할 때

persons arriving at or departing from each such place. The Neutral Nations Supervisory Commission, through its Neutral Nations Inspection Teams, shall conduct supervision and inspection of the rotation of units and personnel authorized above, at the ports of entry enumerated in Paragraph 43 hereof.

d. Cease the introduction into Korea of reinforcing combat aircraft, armored vehicles, weapons, and ammunition; provided, however, that combat aircraft, armored vehicles, weapons, and ammunition which are destroyed, damaged, worn out, or used up during the period of the armistice may be replaced on the basis of piece-for-piece of the same effectiveness and the same type. Such combat aircraft, armored vehicles, weapons, and ammunitions shall be introduced into Korea only through the ports of entry enumerated in Paragraph 43 hereof. In order to justify the requirement for combat aircraft, armored vehicles, weapons, and ammunition to be introduced into Korea for replacement purposes, (b) Refer to UNC Position declared at the 75th MAC Meeting, 21 June '57 reports concerning every incoming shipment of these items shall be made to the Military Armistice Commission and the Neutral Nations Supervisory Commission; such reports shall include statements regarding the disposition of the items being replaced. Items to be replaced which are removed from Korea shall be removed only through the ports of entry enumerated in Paragraph 43 hereof. The Neutral Nations Supervisory Commission, through its Neutral Nations Inspection Teams, shall conduct supervision and inspection of the replacement of combat aircraft, armored vehicles, weapons and ammunition authorized above, at the ports of entry enumerated in Paragraph 43 hereof.

e. Insure that personnel of their respective commands who violate

이 보고는 입경과 출경의 지점 및 매개 지점에서 입경하는 인원과 출경하는 인원의 수자를 포함한다. 중립국감독위원회는 그의 중립국시찰소조를 통하여 본 정전협정 제43항에 열거한 출입항에서 상기의 허가된 부대 및 인원의 윤환을 감독하며 시찰한다.

ㄹ) 한국 경외로부터 증원하는 작전비행기, 장갑차량, 무기 및 탄약을 들여오는 것을 정지한다. 단 정전기간에 파괴, 파손, 손괴 또는 소용된 작전비행기, 장갑차량, 무기 및 탄약은 같은 성능과 같은 유형의 물건을 일(1) 대 일(1)로 교환하는 기초하에서 교체할 수 있다. 이러한 작전비행기, 장갑차량, 무기 및 탄약은 오직 본 정전협정 제43항에 열거한 출입항을 경유하여서만 한국으로 들여올 수 있다. 교체의 목적으로 작전비행기, 장갑차량, 무기 및 탄약을 한국으로 반입할 필요를 확증하기 위하여 이러한 물건의 매차 반입에 관하여 군사정전위원회와 중립국감독위원회에 보고한다.

이 보고 중에서 교체되는 물건의 처리정형을 설명한다. 교체되어 한국으로부터 내어가는 물건은 오직 본 정전협정 제43항에 열거한 출입항을 경유하여서만 내어갈 수 있다.

중립국감독위원회는 그의 중립국시찰소조를 통하여 본 정전협정 제43항에 열거한 출입하에서 상기의 허가된 작전비행기, 장갑차량, 무기 및 탄약의 교체를 감독하여 시찰한다.

ㅁ) 본 정전협정중의 어떠한 규정이든지 위반하는 각

까지 해당 규정에 의해 부과되는 의무의 이행으로부터 벗어날 것임을 선언을 하였습니다. 오늘날 남측과 북측은 정전 당시와는 비교할 수 없을 만큼 새로이 군사적 무장이 많이 되었습니다.

any of the provisions of this Armistice Agreement are adequately punished.

f. In those cases where places of burial are a matter of record and graves are actually found to exist, permit graves registration personnel of the other side to enter, within a definite time limit after this Armistice Agreement becomes effective, the territory of Korea under their military control, for the purpose of proceeding to such graves to recover and evacuate the bodies of the deceased military personnel of that side, including deceased prisoners of war. The specific procedures and the time limit for the performance of the above task shall be determined by the Military Armistice Commission. The Commanders of the opposing sides shall furnish to the other side all available information pertaining to the places of burial of the deceased military personnel of the other side.

g. Afford full protection and all possible assistance and cooperation to the Military Armistice Commission, its Joint Observer Teams, the Neutral Nations Supervisory Commission, and its Neutral Nations Inspection Teams, in the carrying out of their functions and responsibilities hereinafter assigned; and accord to the Neutral Nations Supervisory Commission, and to its Neutral Nations Inspection Teams, full convenience of movement between the headquarters of the Neutral Nations Supervisory Commission and the ports of entry enumerated in Paragraph 43 hereof over main lines of communication agreed upon by both sides(See Map 4), and between the headquarters of the Neutral Nations Supervisory Commission and the places where violations of this Armistice Agreement have been reported to have occurred. In order to prevent unnecessary delays, the use of alternate routes and means of transportation will be permitted whenever the main lines of

자의 지휘하에 있는 인원을 적당히 처벌할 것을 보장한다.

ㅂ) 매장지점이 기록에 있고 분묘가 확실히 존재하고 있다는 것이 판명된 경우에는 본 정전협정이 효력을 발생한 후 일정한 기한내에 그의 군사 통제하에 있는 한국지역에 상대방의 분묘등록 인원이 들어오는 것을 허가하여 이러한 분묘 소재지에 가서 해당측의 이미 죽은 전쟁포로를 포함한 죽은 군사인원의 시체를 발굴하고 또 반출하여 가도록 한다. 상기 사업을 진행하는 구체적 방법과 기한은 군사정전위원회가 이를 결정한다.

적대쌍방 사령관들은 상대방의 죽은 군사인원의 매장지점에 관계되는 얻을 수 있는 일체 재료를 상대방에 제공한다.

ㅅ) 군사정전위원회와 그의 공동감시소조 및 중립국감독위원회와 그의 중립국시찰소조가 하기와 같이 지정한 그들의 직책과 임무를 집행할 때에 충분한 보호 및 일체의 가능한 방조와 협력을 한다.

중립국감독위원회 및 그의 중립국시찰소조가 쌍방이 합의한 주요 교통선을 경유하여 중립국감독위원회 본부와 본 정전협정 제43항에 열거한 출입항간을 왕래할 때와 또 중립국감독위원회 본부와 본 정전협정 위반사건이 발생하였다고 보고된 지점간을 왕래할 때에 충분한 통행상의 편리를 준다.

불필요한 지연을 방지하기 위하여 주요 교통선이 막히든지 통행할 수 없는 경우에는 다른 통로와 수송기재를 사용할 것을 허가한다.

• 중립국감독위원회의 임무는 정전협정 후 한반도에 새로이 군사력 증강을 것을 금지시키는 것이었습니다. 이를 지원하기 위해 20개의 중립국시찰소조를 두고 남북한의 지정된 출입항에 각각 5개 조씩 주재시키고, 나머지 10개 조는 이동시찰소조 임무를 수행하였습니다.(남측 5곳; 인천, 대구, 부산, 강릉, 군산. 북측 5곳; 신의주, 청진, 흥남, 만포, 신안주)

communication are closed or impassable.

h. Provide such logistic support, including communications and transportation facilities, as may be required by the Military Armistice Commission and the Neutral Nations Supervisory Commission and their Teams.

i. Each construct, operate, and maintain a suitable airfield in their respective parts of the Demilitarized Zone in the vicinity of the headquarters of the Military Armistice Commission, for such used as the Commission may determine.

j. Insure that all members and other personnel of the Neutral Nations Supervisory Commission and of the Neutral Nations Repatriation Commission hereinafter establishes shall enjoy the freedom and facilities necessary for the proper exercise of their functions, including privileges, treatment, and immunities equivalent to those ordinarily enjoyed by accredited diplomatic personnel under international usage.

14. This Armistice Agreement shall apply to all opposing ground forces under the military control of either side, which ground forces shall respect the Demilitarized Zone and the area of Korea under the military control of the opposing side.

15. This Armistice Agreement shall apply to all opposing naval forces, which naval forces shall respect the waters contiguous to the Demilitarized Zone and to the land area of Korea under the military control of the opposing side, and shall not engage in blockade of any kind of Korea.

16. This Armistice Agreement shall apply to all opposing air forces,

ㅇ) 군사정전위원회 및 중립국감독위원회와 그 각자에 속하는 소조에 요구되는 통신 및 운수상 편리를 포함한 보급상의 원조를 제공한다.

ㅈ) 군사정전위원회 본부 부근 비무장지대내의 자기측 지역에 각각 한 개의 적당한 비행장을 건설, 관리 및 유지한다. 그 용도는 군사정전위원회가 결정한다.

ㅊ) 중립국감독위원회와 하기와 같이 설립한 중립국송환의원회의 전체 위원 및 기타 인원이 두 자리의 직책을 적당히 집행함에 필요한 자유와 편리를 가지도록 보장한다. 이에는 인가된 외교인원이 국제관례에 따라 통상적으로 향유하는 바와 동등한 특권, 대우 및 면제권을 포함한다.

14. 본 정전협정을 쌍방의 군사통제하에 있는 적대 중의 일체 지상 군사역량에 적용되며 이러한 지상 군사역량은 비무장지대와 상대방의 군사통제하에 있는 한국지역을 존중한다.

15. 본 정전협정은 적대 중의 일체 해상 군사역량에 적용되며 이러한 해상 군사역량은 비무장지대와 상대방의 군사통제하에 있는 한국 육지에 인접한 해면을 존중하며 항구에 대하여 어떠한 종류의 봉쇄도 하지 못한다.

16. 본 정전협정은 적대 중의 일체 공중 군사역량에 적용

• 유엔사측은 판문점내 헬기 비행 권한 행사를 위한 월1회 판문점 헬기 비행훈련(JSA-판문점 H-128)을 하고 있습니다.

비행훈련

• "항구에 대하여 어떠한 종류의 봉쇄도 하지 못한다"는 조항은 공산군측의 해안봉쇄 우려를 해소시키기 위해 포함한 조항입니다.
정전협정 당시 지상의 군사분계선은 설정되었으나, 해상에서의 군사분계선은 공산군측과 유엔군측의 의견이 달라 합의되지 않았습니

which air forces shall respect the air space over the Demilitarized Zone and over the area of Korea under the military control of the opposing side, and over the waters contiguous to both.

17. Responsibility for compliance with and enforcement of the terms and provisions of this Armistice Agreement is that of the signatories hereto and their successors in command. The Commanders of the opposing sides shall establish within their respective commands all measures and procedures necessary to insure complete compliance with all of the provisions hereof by all elements of their commands. They shall actively cooperate with one another and with the Military Armistice Commission and the Neutral Nations Supervisory Commission in requiring observance of both the letter and the spirit of all of the provisions of this Armistice Agreement.

18. The costs of the operations of the Military Armistice Commission and of the Neutral Nations Supervisory Commission and of their Teams shall be shared equally by the two opposing sides.

B. MILITARY ARMISTICE COMMISSION

1. COMPOSITION

19. A Military Armistice Commission is hereby established.

20. The Military Armistice Commission shall be composed of ten(10) senior officers, five(5) of whom shall be appointed by the Commander-in-Chief, United Nations Command, and five(5) of

되며 이러한 공중 군사역량은 비무장지대와 상대방의 군사통제하에 있는 한국 지역 및 이양 지역에 인접한 해면의 상공을 존중한다.

17. 본 정전협정의 조항과 규정을 준수하며 집행하는 책임은 본 정전협정에 조인한 자와 그의 후임 사령관에게 속한다.
적대 쌍방 사령관들은 각각 그들의 지휘하에 있는 군대내에서 일체의 필요한 조치와 방법을 취함으로써 그 모든 소속 부대 및 인원이 본 정전협정의 전체 규정을 철저히 준수하는 것을 보장한다.
적대 쌍방 사령관들은 호상 적극 협력하여 군사정전위원회 및 중립국감독위원회와 적극 협력함으로써 본 정전협정 전체 규정의 문구와 정신을 준수하도록 한다.

18. 군사정전위원회와 중립국감독위원회 및 그 각자에 속하는 소조의 사업비용은 적대 쌍방이 균등하게 부담한다.

나. 군사정전위원회

1. **구성**

19. 군사정전위원회를 설립한다.

20. 군사정전위원회는 십(10) 명의 고급장교로 구성하되 그 중의 오(5) 명은 국제연합군 총사령관이 이를 임명하며 그 중의 오(5) 명은 조선인민군 최고사령관과 중

다. 제공권과 제해권을 장악하고 있던 유엔군측은 당시 국제적으로 통용되는 3해리(NM) 영해를 주장하였고, 공산군측은 유엔군측의 해안봉쇄를 우려하여 12해리(NM) 영해를 주장하여 쌍방의 의견 차이가 컸기 때문입니다. 상기 조항을 포함시켜 공산군측의 해안봉쇄 우려를 해소시키면서 해상에서의 군사분계선을 협의하였으나 합의에 이르지 못했습니다.

• 군사정전위원회 유엔군측은 미국, 한국, 영국, 그리고

whom shall be appointed jointly by the Supreme Commander of the Korean People's Army and the Commander of the Chinese People's Volunteers. Of the ten members, three(3) from each side shall be of general or flag rank. The two(2) remaining members on each side may be major generals, brigadier generals, colonels, or their equivalents.

21. Members of the Military Armistice Commission shall be permitted to use staff assistants as required.

22. The Military Armistice Commission shall be provided with the necessary administrative personnel to establish a Secretariat charged with assisting the Commission by performing record-keeping, secretariat, interpreting, and such other functions as the Commission may assign to it. Each side shall appoint to the Secretariat a Secretary and an Assistant Secretary and such clerical and specialized personnel as required by the Secretariat. Records shall be kept in English, Korean, and Chinese, all of which shall be equally authentic.

23. a. The Military Armistice Commission shall be initially provided with and assisted by ten(10) Joint Observer Teams, which number may be reduced by agreement of the senior members of both sides on the Military Armistice Commission.

b. Each Joint Observer Teams shall be composed of not less than four(4) nor more than six(6) officers of field grade, half of whom shall be appointed by the Commander-in-Chief, United Nations Command, and half of whom shall be appointed jointly by the Supreme Commander of the Korean People's Army and the

국인민지원군 사령원이 공동으로 이를 임명한다. 위원 십(10) 명중에서 각방의 삼(3) 명은 장급에 속하여야 하며 각방의 나머지 이(2) 명은 소장, 준장, 대령 혹은 그와 동급인 자로 할 수 있다.

참전국 연락장교단 대표들을 임명하고 있으며 수석대표는 정전협정 체결 이후 미군 장성이 임명되어 오다가 1991년 3월 25일 제57대 수석대표로 황원탁 소장이 임명된 이후 한국군 장성이 임명되어 오고 있습니다.

21. 군사정전위원회의 위원은 그 필요에 따라 참 보조인원을 사용할 수 있다.

22. 군사정전위원회는 필요한 행정인원을 배치하여 비서처를 설치하되 그 임무는 동 위원회의 기록, 서기, 통역 및 동 위원회가 지정하는 기타의 직책의 집행을 협조하는 것이다. 쌍방은 각기 비서처에 비서장 일(1) 명 보조 비서장 일(1) 명 및 비서처에 필요한 서기 및 전문 기술인원을 임명한다. 기록은 영문, 한국문 및 중국문으로 작성하되 세 가지 글은 동등한 효력을 가진다.

• 현재 비서장은 미측 대령 1명, 부비서장은 한측 대령 1명, 비서처 참모인원은 한국, 미국, 캐나다, 뉴질랜드, 콜롬비아, 덴마크 등으로 구성되어 있습니다.

23. ㄱ) 군사정전위원회는 처음에는 십(10) 개의 공동감시소조를 두어 그 협조를 받는다. 소조의 수는 군사정전위원회의 쌍방 수석위원의 합의를 거쳐 감소할 수 있다.
ㄴ) 매개의 공동감시소조는 사(4) 명 내지 육(6) 명의 영급 장교로 구성하되 그 중의 반수는 국제연합군 총사령관이 이를 임명하여 그 중의 반수는 조선인민군 최고사령관과 중국인민지원군 사령원이 공동으로 이를 임명한다. 공동감시소조의 사업상 필요한 운전수,

• 군사정전위원회 공동감시소조는 1953년에 10개 소조, 1954년에 6개 소조, 1955년에는 5개 소조로 축소되었으며, 당시 담당구역은 제1조는 한강하구, 제2조는 표식물 0001호부터 0343호, 제3조는 제2조 동단부터 표식물 0710까지, 제4조는 제3조의 동단부터 표식물 제0982호까지, 제5조는 제4조의 동단부터 표식물 1292호까지로 조정되었습니다.

Commander of the Chinese People's Volunteers. Additional personnel such as drivers, clerks, and interpreters shall be furnished by each side as required for the functioning of the Joint Observer Teams.

2. FUNCTIONS AND AUTHORITY

24. The general mission of the Military Armistice Commission shall be to supervise the implementation of this Armistice Agreement and to settle through negotiations any violations of this Armistice Agreement.

25. The Military Armistice Commission shall:

a. Locate its headquarters in the vicinity of PANMUNJOM(37°57' 29" N, 126°40' 00" E). The Military Armistice Commission may relocate its headquarters at another point within the Demilitarized Zone by agreement of the senior members of both sides on the Commission

b. Operate as a joint organization without a chairman.

c. Adopt such rules of procedure as it may, from time to time, deem necessary.

d. Supervised the carrying out of the provisions of this Armistice Agreement pertaining to the Demilitarized Zone and to the Han River Estuary.

e. Direct the operations of the Joint Observer Teams.

f. Settle through negotiations any violations of this Armistice Agreement.

g. Transmit immediately to the Commanders of the opposing sides

서기 통역 등의 부속인원은 쌍방이 이를 제공한다.

공동감시소조 참여에 대한 북한군측의 미온적인 태도로 인해 1968년 유엔군사령관은 통제지역내에서 유엔사 특별조사반(SIT)에 의한 조사의 활용을 지시하였습니다.
이후 조금씩 부실화되기 시작하여 1994년부터는 거의 기능이 마비되었습니다.

2. 직책과 권한

24. 군사정전위원회의 전반적 임무는 본 정전협정의 실시를 감독하며 본 정전협정의 어떠한 위반사건이든지 협의하여 처리하는 것이다.

25. 군사정전위원회는

ㄱ) 본부를 판문점(북위 37도57분29초, 동경 126도40분00초) 부근에 설치한다. 군사정전위원회는 동 위원회의 쌍방 수석위원의 합의를 거쳐 그 본부를 비무장지대내의 다른 한 지점에 이설할 수 있다.

ㄴ) 공동기구로서 사업을 진행하며 의장을 두지 않는다.

ㄷ) 그가 수시로 필요하다고 인정하는 절차 규정을 채택한다.

ㄹ) 본 정전협정중 비무장지대와 한강 하구에 관한 각 규정의 집행을 감독한다.

ㅁ) 공동감시소조의 사업을 지도한다.

ㅂ) 본 정전협정의 어떠한 위반사건이든지 협의하여 처리한다.

ㅅ) 중립국감독위원회로부터 받은 본 정전협정 위반

• 정전 이후 양측은 정전협정

all reports of investigations of violations of this Armistice Agreement and all other reports and records of proceedings received from the Neutral Nations Supervisory Commission.

h. Give general supervision and direction to the activities of the Committee for Repatriation of Prisoners of War and the Committee for Assisting the Return of Displaced Civilians, hereinafter established.

i. Act as an intermediary in transmitting communications between the Commanders of the opposing sides; provided, however, that the foregoing shall not be construed to preclude the Commanders of both sides from communicating with each other by any other means which they may desire to employ.

j. Provide credentials and distinctive insignia for its staff and its Joint Observer Teams, and a distinctive marking for all vehicles, aircraft, and vessels, used in the performance of its mission.

26. The mission of the Joint Observer Teams shall be to assist the Military Armistice Commission in supervising the carrying out of the provisions of this Armistice Agreement pertaining to the Demilitarized Zone and to the Han River Estuary.

27. The Military Armistice Commission, or the senior member of either side thereof, is authorized to dispatch Joint Observer Teams to investigate violations of this Armistice Agreement reported to have occurred in the Demilitarized Zone or in the Han River Estuary; provided, however, that not more than one half of the Joint Observer Teams which have not been dispatched by the Military Armistice Commission may be dispatched at any one time by the senior member of either side on the Commission.

사건에 관한 일체 조사 보고 및 일체 기타 보고와 회의기록은 즉시로 적대쌍방 사령관들에게 이를 전달한다.

ㅇ) 하기한 바와 같이 설립한 전쟁포로송환위원회와 실향사민귀향협조위원회의 사업을 전반적으로 감독하며 지도한다.

ㅈ) 적대쌍방 사령관간에 통신을 전달하는 중개 역할을 담당한다. 단 상기의 규정은 쌍방 사령관들이 사용하고자 하는 어떠한 다른 방법을 사용하여 호상 통신을 전달하는 것을 배제하는 것으로 해석할 수 없다.

ㅊ) 그의 공작인원과 그의 공동감시소조의 증명 문건 및 휘장 또 그 임무 집행시에 사용하는 일체의 차량, 비행기 및 선박의 식별 표식을 발급한다.

26. 공동감시소조의 임무는 군사정전위원회가 본 정전협정 중의 비무장지대 및 한강 하구에 관한 각 규정의 집행을 감독함을 협조하는 것이다.

27. 군사정전위원회 또는 그 중의 어느 일방의 수석위원은 공동감시소조를 파견하여 비무장지대나 한강 하구에서 발생하였다고 보고된 본 정전협정 위반사건을 조사할 권한을 가진다.
단 동 위원회 중의 어느 일방의 수석위원이든지 언제나 군사정전위원회가 아직 파견하지 않은 공동감시소조의 반수 이상을 파견할 수 없다.

위반현황 통계를 매월 판문점에서 상호 통보하고 교환해 왔으나, 북한은 1994년 군사정전위원회의 무실화를 주장하면서 통계교환을 일방적으로 중단하였고 유엔군측에서도 공식통계를 작성하지 않고 있습니다.

• 전쟁포로송환위원회는 각각 영관급장교 3명씩 6명으로 구성되며, 전쟁포로 송환에 관계되는 계획을 조정하며 실시하는 것을 감독합니다.
실향사민 귀향협조위원회는 영급장교 2명씩 총4명으로 구성되며 일반인의 비무장지대내 귀향계획을 조절하며 쌍방의 집행을 감독합니다.

• 비무장지대와 한강하구지역을 제외한 모든 곳은 중립국감독위원회가 정전집행을 감독하고 결과를 군사정전위원회에 보고합니다.

28. The Military Armistice Commission, or the senior member of either side thereof, is authorized to request the Neutral Nations Supervisory Commission to conduct special observations and inspections at places outside the Demilitarized Zone where violations of this Armistice Agreement have been reported to have occurred.

29. When the Military Armistice Commission determines that a violation of this Armistice Agreement has occurred, it shall immediately report such violation to the Commanders of the opposing sides.

30. When the Military Armistice Commission determines that a violation of this Armistice Agreement has been corrected to its satisfaction, it shall so report to the Commanders of the opposing sides.

3. GENERAL

31. The Military Armistice Commission shall meet daily. Recesses of not to exceed seven(7) days may be agreed upon by the senior members of both sides; provided, that such recesses may be terminated on twenty-four(24) hour notice by the senior member of either side.

32. Copies of the record of the proceedings of all meetings of the Military Armistice Commission shall be forwarded to the Commanders of the opposing sides as soon as possible after each meeting.

28. 군사정전위원회 또는 동 위원회의 어느 일방의 수석 위원은 중립국감독위원회에 요청하여 본 정전협정 위반사건이 발생하였다고 보고된 비무장지대 이외의 지점에 가서 특별한 감시와 시찰을 행할 권한을 가진다.

29. 군사정전위원회가 본 정전협정 위반사건이 발생하였다고 확정한 때에는 즉시로 그 위반사건을 적대쌍방 사령관들에게 보고한다.

30. 군사정전위원회가 본 정전협정의 어떠한 위반사건이 만족하게 시정되었다고 확정한 때에는 이를 적대쌍방 사령관들에게 보고한다.

3. 총칙

31. 군사정전위원회는 매일 회의를 연다. 쌍방의 수석위원은 합의하여 칠(7) 일을 넘지 않는 휴회를 할 수 있다. 단 어느 일방의 수석위원이든지 이십사(24) 시간 전의 통고로써 이 휴회를 끝낼 수 있다.

32. 군사정전위원회의 일체 회의기록의 부분은 매번 회의 후 될 수 있는 대로 속히 적대 쌍방 사령관들에게 송부한다.

• 본회의는 정전협정의 이행을 감독하고 주요 정전협정 위반사항에 대해 협의 · 처리해 왔으며, 1953년 7월 28일 제1차 회의를 시작한 이후 1991년 2월까지 총 459차례 개최되었습니다. 그러나 1991년 3월 25일 유엔사측 수석대표로 한국군 장성이 임명된 후 북한측은 한국군 수석대표 불인정 및 본회의 참석을 거부하였고, 공산군측 대표단의 철수조치로 인해 그 기능이 마비되었습니다. 그 이후 남북간 우발적

33. The Joint Observer Teams shall make periodic reports to the Military Armistice Commission as required by the Commission and, in addition, shall make such special reports as may be deemed necessary by them, or as may be required by the Commission.

34. The Military Armistice Commission shall maintain duplicate files of the reports and records of proceedings required by this Armistice Agreement. The Commission is authorized to maintain duplicate files of such other reports, records, etc., as may be necessary in the conduct of its business. Upon eventual dissolution of the commission, one set of the above files shall be turned over to each side.

35. The Military Armistice Commission may make recommendations to the Commanders of the opposing sides with respect to amendments or additions to this Armistice Agreement. Such recommended changes should generally be those designed to insure a more effective armistice.

C. NEUTRAL NATIONS SUPERVISORY COMMISSION

1. COMPOSITION

36. A Neutral Nations Supervisory Commission is hereby established.

37. The Neutral Nations Supervisory Commission shall be composed of four(4) senior officers, two(2) of whom shall be appointed by neutral nations nominated by the Commander-in-Chief, United Nations Command, namely, SWEDEN and SWITZERLAND, and

33. 공동감시소조는 군사정전위원회에 동 위원회가 요구하는 정기 보고를 제출하며 또 이 소조들이 필요하다고 인정하거나 또는 동 위원회가 요구하는 특별보고를 제출한다.

충돌 발생시 평화적 해결을 위한 필요성이 제기되어 1998년 6월부터 유엔사 및 조선인민군 장성급회담으로 대체되어 정전문제를 협의 · 처리해 오고 있습니다.

34. 군사정전위원회는 본 정전협정에 규정한 보고 및 회의 기록의 문건철 두 벌을 보관한다. 동 위원회는 그 사업 진행에 필요한 기타의 보고 기록 등의 문건철 두 벌을 보관할 권한을 가진다. 동 위원회의 최후 해산시에는 상기 문건철을 쌍방에 각 한 벌씩 나누어 준다.

35. 군사정전위원회는 적대 쌍방 사령관들에게 본 정전협정의 수정 또는 증보에 대한 건의를 제출할 수 있다. 이러한 개정 건의는 일반적으로 더 유효한 정전을 보장할 것을 목적으로 하는 것이어야 한다.

다. 중립국감독위원회

1. 구성

36. 중립국감독위원회를 설립한다.

37. 중립국감독위원회는 사(4) 명의 고급장교로 구성하되 그중의 이(2) 명은 국제연합군 총사령관이 지명한 중립국 즉 서전 및 서서가 이를 임명하며 그 중의 이(2) 명은 조선인민군 최고사령관과 중국인민지원군

• 중립국감독위원회의 임무는 정전협정 후 한반도에 새로이 군사력 증강을 금지시키는 것입니다.

two(2) of whom shall be appointed by neutral nations nominated jointly by the Supreme Commander of the Korean People's Army and the Commander of the Chinese People's Volunteers, namely, POLAND and CZECHOSLOVAKIA. The term "neutral nations" as herein used is defined as those nations whose combatant forces have not participated in the hostilities in Korea. Members appointed to the Commission may be from the armed forces of the appointing nations. Each member shall designate an alternate member to attend those meetings which for any reason the principal member is unable to attend. Such alternate members shall be of the same nationality as their principals. The Neutral Nations Supervisory Commission may take action whenever the number of members present from the neutral nations nominated by one side is equal to the number of members present from the neutral nations nominated by the other side.

38. Members of the Neutral Nations Supervisory Commission shall be permitted to use staff assistants furnished by he neutral nations as required. These staff assistants may be appointed as alternate members of the Commission.

39. The neutral nations shall be requested to furnish the Neutral Nations Supervisory Commission with the necessary administrative personnel to establish a Secretariat charged with assisting the Commission by performing necessary record-keeping, secretarial, interpreting, and such other functions as the Commission may assign to it.

40. a. The Neutral Nations Supervisory Commission shall be initially

사령원이 공동으로 지명한 중립국 즉 파란 및 체코슬로바키아가 이를 임명한다. 본 정전협정에서 쓴 '중립국' 이라는 어의 정의는 그 전투부대가 한국에서의 적대행위에 참가하지 않은 국가를 말하는 것이다. 동 위원회에 임명되는 위원은 임명하는 국가의 무장부대로부터 파견될 수 있다. 매개 위원은 후보위원 일(1) 명을 지정하여 그 정위원이 어떤 이유로 출석할 수 없게 되는 회의에 출석하게 된다. 이러한 후보위원은 그 정위원과 동일한 국적에 속한다. 일방이 지명한 중립국 위원의 출석자수와 다른 일방이 지명한 중립국위원의 출석자수가 같을 때에는 중립국감독위원회는 곧 행동을 취할 수 있다.

38. 중립국감독위원회의 위원은 그 필요에 따라 각기 해당 중립국가가 제공한 참 보조인원을 사용할 수 있다. 이러한 참 보조인원은 본 위원회의 후보인원으로 임명될 수 있다.

39. 중립국감독위원회에 필요한 행정인원을 제공하도록 중립국에 요청하여 비서처를 설치하되 그 임무는 동 위원회에 필요한 기록, 서기, 통역 및 동 위원회가 지정하는 기타의 직책의 집행을 협조하는 것이다.

40. ㄱ) 중립국감독위원회는 처음에는 이십(20) 개의 중

• 서전 : 스웨덴
서서 : 스위스
파란 : 폴란드

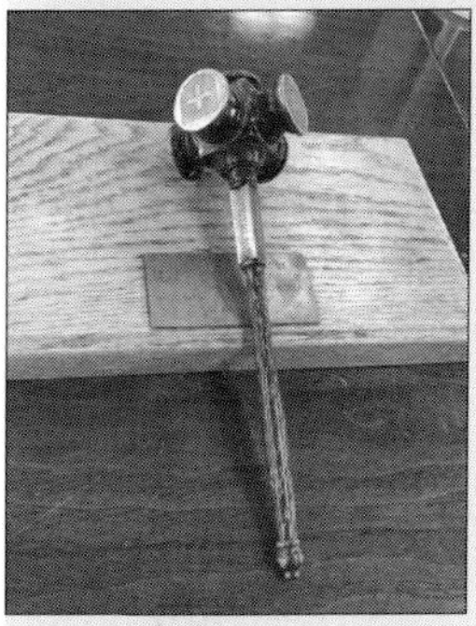

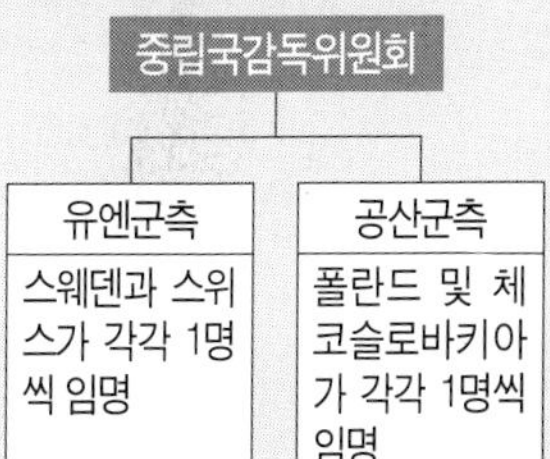

provided with, and assisted by, twenty(20) Neutral Nations Inspection Teams, which number may be reduced by agreement of the senior members of both sides on the Military Armistice Commission. The Neutral Nations Inspection Teams shall be responsible to, shall report to, and shall be subject to the direction of, the Neutral Nations Supervisory Commission only.

b. Each Neutral Nations Inspection Team shall be composed of not less than four(4) officers, preferably of field grade, half of whom shall be from the neutral nations nominated by the Commander-in-Chief, United Nations Command, and half of whom shall be from the neutral nations nominated jointly by the Supreme Commander of the Korean People's Army and the Commander of the Chinese People's Volunteers. Members appointed to the Neutral Nations Inspection Teams may be from the armed forces of the appointing nations. In order to facilitate the functioning of the Teams, sub-teams composed of not less than two(2) members, one of whom shall be from a neutral nation nominated by the Commander-in-Chief, United Nations Command, and one of whom shall be from a neutral nation nominated jointly by the Supreme Commander of the Korean People's Army and the Commander of the Chinese People's Volunteers, may be formed as circumstances require. Additional personnel such as drivers, clerks, interpreters, and communications personnel, and such equipment as may be required by the Teams to perform their mission, shall be furnished by the Commander of each side, as required, in the Demilitarized Zone and in the territory under his military control. The Neutral Nations Supervisory Commission may provide itself and the Neutral Nations Inspection Teams with such of the above personnel and equipment of its own as it may desire; provided, however, that such personnel shall be

립국 시찰소조를 두어 그 협조를 받는다. 소조의 수는 군사정전위원회의 쌍방 수석위원의 합의를 거쳐 감소할 수 있다. 중립국시찰소조는 오직 중립국감독위원회에 대하여서만 책임을 지며, 그에 보고하며 또 그 지도를 받는다.

ㄴ) 매개 중립국시찰소조는 최소 사(4) 명의 장교로 구성하되 이 장교는 영급으로 하는 것이 적당하며 그 중의 반수는 국제연합군 총사령관이 지명한 중립국에서 내고 그 중의 반수는 조선인민군 최고사령관과 중국인민지원군 사령원이 공동으로 지명한 중립국에서 낸다. 중립국시찰소조에 임명되는 조원은 임명하는 국가의 무장부대에서 이를 낼 수 있다. 각 소조의 직책집행을 편리하게 하기 위하여 정황의 요구에 따라 최소 이(2) 명의 조원으로 구성하는 분조를 설치할 수 있다. 그 두 조원 중의 일(1) 명은 국제연합군 총사령관이 지명한 중립국에서 내며 일(1) 명은 조선인민군 최고사령관과 중국인민지원군 사령원이 공동으로 지명한 중립국에서 낸다. 운전수, 서기, 통역, 통신원과 같은 부속인원 및 각 소조의 임무 집행에 필요한 비품은 각방 사령관이 비무장지대내 및 자기측 군사통제 지역내에서 수요에 따라 이를 공급한다. 중립국감독위원회는 동 위원회 자체와 중립국시찰소조들에 그가 요망하는 상기의 인원 및 비품을 제공할 수 있다. 단 이러한 인원은 중립국감독위원회를 구성한 그 중립국의 인원이어야 한다.

• 중립국감독위원회의 규모는 1953년 말에는 유엔군측이 각각 100여 명, 공산군측이 각각 300여 명 규모였으나, 1956년 이후 기능과 활동범위가 감소됨에 따라 각 국가별 10여 명 수준으로 급감하였고 현재는 유엔군측만 각각 5명씩으로 임무를 수행하고 있습니다.

personnel of the same neutral nations of which the Neutral Nations Supervisory Commission is composed.

2. FUNCTIONS AND AUTHORITY

41. The mission of the Neutral Nations Supervisory Commission shall be to carry out the functions of supervision, observation, inspection, and investigation, as stipulated in Sub-paragraphs 13c and 13d and Paragraph 28 hereof, and to report the results of such supervision, observation, inspection, and investigation to the Military Armistice Commission.

42. The Neutral Nations Supervisory Commission shall:

a. Locate its headquarters in proximity to the headquarters of the Military Armistice Commission.

b. Adopt such rules of procedure as it may, from time to time, deem necessary.

c. Conduct, through its members and its Neutral Nations Inspection Teams, the supervision and inspection provided for in Sub-paragraphs 13c and 13d of this Armistice Agreement at the ports of entry enumerated in Paragraph 43 hereof, and the special observations and inspections provided for in Paragraph 28 hereof at those places where violations of this Armistice Agreement have been reported to have occurred. The inspection of combat aircraft, armored vehicles, weapons, and ammunition by the Neutral Nations Inspection Teams shall be such as to enable them to properly insure that reinforcing combat aircraft, armored vehicles, weapons, and ammunition are not being introduced into Korea; but this shall not be construed as authorizing inspections or

2. 직책과 권한

41. 중립국감독위원회의 임무는 본 정전협정 제13항 ㄷ, 제13항 ㄹ 및 제28항에 규정한 감독, 감시, 시찰 및 조사의 직책을 집행하며 이러한 감독, 감시, 시찰 및 조사의 결과를 군사정전위원회에 보고하는 것이다.

42. 중립국감독위원회는

ㄱ) 본부를 군사정전위원회 본부의 부근에 설치한다.

ㄴ) 그가 수시로 필요하다고 인정하는 절차 규정을 채택한다.

ㄷ) 그 위원 및 그 중립국시찰소조를 통하여 본 정전협정 제43항에 열거한 출입항에서 본 정전협정 제13항 ㄷ, 제13항 ㄹ에 규정한 감독과 시찰을 진행하며 또 본 정전협정 위반사건이 발생하였다고 보고된 지점에서 본 정전협정 제28항에 규정한 특별 감시와 시찰을 진행한다. 작전비행기, 장갑차량, 무기 및 탄약에 대한 중립국시찰소조의 시찰은 소조로 하여금 증원하는 작전비행기, 장갑차량, 무기 및 탄약을 한국으로 들여옴이 없도록 확실히 보장할 수 있게 한다.
단 이 규정은 어떠한 작전비행기, 장갑차량, 무기 또는 탄약의 어떠한 비밀 설계 또는 특점을 시찰 또는 검사할 권한을 주는 것으로 해석할 수 없다.

• 정전협정 당시 유엔군측은 군사정전위원회가 한반도 모든 지역에 대한 감독을 하도록 계획하였으나 공산군측은 북한군의 무력증강에 지장을 주지 않도록 군사정전위원회는 비무장지대와 한강하구수역을 그리고 나머지 지역은 중립국감독위원회가 감독하도록 제안하였습니다. 공산군측은 당시 소련의 위성국이었던 폴란드와 체코슬로바키아를 추천하였고 유엔사측은 중립국인 스위스와 스웨덴을 추천하였습니다.

examinations of any secret designs or characteristics of any combat aircraft, armored vehicle, weapon, or ammunition.

d. Direct and supervise the operations of the Neutral Nations Inspection Teams.

e. Station five(5) Neutral Nations Inspection Teams at the ports of entry enumerated in Paragraph 43 hereof located in the territory under the military control of the Commander-in-Chief, United Nations Command; and five(5) Neutral Nations Inspection Teams at the ports of entry enumerated in Paragraph 43 hereof located in the territory under the military control of the Supreme Commander of the Korean People's Army and the Commander of the Chinese People's Volunteers; and establish initially ten(10) mobile Neutral Nations Inspection Teams in reserve, stationed in the general vicinity of the headquarters of the Neutral Nations Supervisory Commission, which number may be reduced by agreement of the senior members of both sides on the Military Armistice Commission. Not more than half of the mobile Neutral Nations Inspection Teams shall be dispatched at any one time in accordance with requests of the senior member of either side on the Military Armistice Commission.

f. Subject to the provisions of the preceding Sub-paragraph, conduct without delay investigations of reported violations of this Armistice Agreement, including such investigations of reported violations of this Armistice Agreement as may be requested by the Military Armistice Commission or by the senior member of either side on the Commission.

g. Provide credentials and distinctive insignia for its staff and its Neutral Nations Inspection Teams, and a distinctive marking for all vehicles, aircraft, and vessels, used in the performance of its mission.

ㄹ) 중립국시찰소조의 사업을 지도하며 감독한다.

ㅁ) 국제연합군 총사령관의 군사통제지역내에 있는 본 정전협정 제43항에 열거한 출입항에 오(5) 개의 중립국시찰소조를 주재시키며 조선인민군 최고사령관과 중국인민지원군 사령원의 군사통제지역내에 있는 본 정전협정 제43항에 열거한 출입항에 오(5) 개의 중립국시찰소조를 주재시킨다. 처음에는 따로 십(10) 개의 중립국이동시찰소조를 후비로 설치하되 중립국감독위원회 본부 부근에 주재시킨다. 그 수는 군사정전위원회의 쌍방 수석위원의 합의를 거쳐 감소시킬 수 있다. 중립국이동시찰소조 중 군사정전위원회의 어느 일방 수석위원의 요청에 응하여 파견하는 소조는 언제나 그 반수를 초과할 수 없다.

• 출입항만과 공항
북측 : 신의주, 청진, 흥남, 만포, 신안주
남측 : 인천, 강릉, 군산, 대구, 부산

ㅂ) 보고된 본 정전협정 위반사건을 전 규정의 범위내에서 지체없이 조사한다. 이에는 군사정전위원회 또는 동 위원회 중의 어느 일방 수석위원이 요청하는 보고된 본 정전협정 위반사건에 대한 조사를 포함한다.

ㅅ) 그의 공작인원과 그의 중립국시찰소조의 증명문건 및 휘장 또 그 임무 집행시에 사용하는 일체 차량, 비행기 및 선박의 식별 표식을 발급한다.

43. Neutral Nations Inspection Teams shall be stationed at the following ports of entry:

Territory under the military control of the United Nations Command	Territory under the military control of the Korean People's Army and the Chinese People's Volunteers
INCHON (37°28' N, 126°38' E)	SINUIJU (40°06' N, 124°24' E)
TAEGU (35°52' N, 128°36' E)	CHONGJIN (41°46' N, 129°49' E)
PUSAN (35°06' N, 129°02' E)	HUNGNAM (39°50' N, 127°37' E)
KANGNUNG (37°45' N, 128°54' E)	MANPO (41°09' N, 126°18' E)
KUNSAN (35°59' N, 126°43' E)	SINANJU (39°36' N, 125°36' E)

These Neutral Nations Inspection Teams shall be accorded full convenience of movement within the areas and over the routes of communication set forth on the attached map(Map 5).

3. GENERAL

44. The Neutral Nations Supervisory Commission shall meet daily. Recesses of not to exceed seven(7) days may be agreed upon by the members of the Neutral Nations Supervisory Commission; provided, that such recesses may be terminated on twenty-four(24) hour notice by any member.

43. 중립국시찰소조는 하기한 각 출입항에 주재한다.

국제연합군의 군사통제지역,	조선인민군과 중국인민지원군의 군사통제지역
인천	신의주
(북위 37도 28분, 동경 126도 38분)	(북위 40도 06분, 동경 124도 24분)
대구	청진
(북위 35도 52분, 동경 128도 36분)	(북위 41도 46분, 동경 129도 49분)
부산	흥남
(북위 35도 06분, 동경 129도 02분)	(북위 39도 50분, 동경 127도 37분)
강릉	만포
(북위 37도 45분, 동경 128도 54분)	(북위 41도 09분, 동경 126도 18분)
군산	신안주
(북위 35도 59분, 동경 126도 43분)	(북위 39도 36분, 동경 125도 36분)

이 중립국시찰소조들은 첨부한 지도에 표시한 지역내와 교통선에서 통행상 충분한 편리를 받는다.(지도5－본서 pp.205~215 참조)

3. 총칙

44. 중립국감독위원회는 매일 회의를 연다.
중립국감독위원회 위원은 합의하여 칠(7) 일을 넘지 않는 휴회를 할 수 있다. 단 어느 위원이든지 이십사(24) 시간 전의 통고로써 이 휴회를 끝낼 수 있다.

• 유엔군측 중립국감독위원회는 현재 매주 1회 회의를 하고 있습니다.

45. Copies of the record of the proceedings of all meetings of the Neutral Nations Supervisory Commission shall be forwarded to the Military Armistice Commission as soon as possible after each meeting. Records shall be kept in English, Korean, and Chinese.

46. The Neutral Nations Inspection Teams shall make periodic reports concerning the results of their supervision, observations, inspections, and investigations to the Neutral Nations Supervisory Commission as required by the Commission and, in addition, shall make such special reports as may be deemed necessary by them, or as may be required by the Commission. Reports shall be submitted by a team as a whole, but may also be submitted by one or more individual members thereof; provided, that the reports submitted by one or more individual members thereof shall be considered as informational only.

47. Copies of the reports made by the Neutral Nations Inspection Teams shall be forwarded to the Military Armistice Commission by the Neutral Nations Supervisory Commission without delay and in the language in which received. They shall not be delayed by the process of translation or evaluation. The Neutral Nations Supervisory Commission shall evaluate such reports at the earliest practicable time and shall forward their findings to the Military Armistice Commission as a matter of priority. The Military Armistice Commission shall not take final action with regard to any such report until the evaluation thereof has been received from the Neutral Nations Supervisory Commission. Members of the Neutral Nations Supervisory Commission and of its Teams shall be subject to appearance before the Military Armistice Commission, at the

45. 중립국감독위원회의 일체 회의기록의 부본은 매번 회의 후 될 수 있는 대로 속히 군사정전위원회에 송부한다. 기록은 영문, 한국문 및 중국문으로 작성한다.

46. 중립국시찰소조는 그의 감독, 감시, 시찰 및 조사의 결과에 관하여 중립국감독위원회가 요구하는 정기보고를 동 위원회에 제출하며 또 이 소조들이 필요하다고 인정하거나 동 위원회가 요구하는 특별보고를 제출한다. 보고는 소조 총체가 이를 제출한다. 단 그 소조의 개별적 조원 일(1)명 또는 수명이 이를 제출할 수도 있다. 개별적 조원 일(1)명 또는 수명이 제출한 보고는 다만 참고적 보고로 간주한다.

• 공산군측 중립국시찰소조는 남한지역에서 간첩행위와 같은 일들을 하면서 유엔군측 중립국시찰소조의 북한지역내 활동을 고의적으로 방해하였습니다. 이에 유엔군측은 1956년 5월 군정위 제70차 본회의에서 20개 중립국시찰소조의 기능을 잠정 중지할 것을 선언하고 남한에 파견된 공산군측 인원을 철수하도록 하였습니다. 그리고 1957년 6월 군정위 제75차 본회의에서 북측의 다량의 군사장비 도입에 대한 정전협정 위반행위를 지적하고 북측이 한국으로의 무기 반입을 금지한 정전협정 제2조 13항의 준수의사를 표시할 때까지 남측도 의무 이행으로부터 벗어날 것임을 선언하였습니다.

47. 중립국감독위원회는 중립국시찰소조가 제출한 보고의 부분을 그가 접수한 보고에 사용된 글로써 지체 없이 군사정전위원회에 송부한다. 이러한 보고는 번역 또는 심의 결정 수속 때문에 지체시킬 수 없다. 중립국감독위원회는 실제 가능한 속히 이러한 보고를 심의 결정하며 그의 판정서를 우선 군사정전위원회에 송부한다. 중립국감독위원회의 해당 심의 결정을 접수하기 전에는 군사정전위원회는 이런 어떠한 보고에 대하여서도 최후적 행동을 취하지 못한다.

군사정전위원회의 어느 일방 수석위원의 요청이 있을 때에는 중립국감독위원회의 위원과 그 소조의 조원은 곧 군사정전위원회에 참석하여 제출된 어떠한 보고에 대하여서든지 설명한다.

request of the senior member of either side on the Military Armistice Commission, for clarification of any report submitted.

48. The Neutral Nations Supervisory Commission shall maintain duplicate files of the reports and records of proceedings required by this Armistice Agreement. The Commission is authorized to maintain duplicate files of such other reports, records, etc., as may be necessary in the conduct of its business. Upon eventual dissolution of the Commission, one set of the above files shall be turned over to each side.

49. The Neutral Nations Supervisory Commission may make recommendations to the Military Armistice Commission with respect to amendments or additions to this Armistice Agreement. Such recommended changes should generally be those designed to insure a more effective armistice.

50. The Neutral Nations Supervisory Commission, or any member thereof, shall be authorized to communicate with any member of the Military Armistice Commission.

ARTICLE III
ARRANGEMENTS RELATING TO PRISONERS OF WAR

51. The release and repatriation of all prisoners of war held in the custody of each side at the time this Armistice Agreement becomes effective shall be effected in conformity with the following provisions agreed upon by both sides prior to the signing of this

48. 중립국감독위원회는 본 정전협정이 규정하는 보고 및 회의기록의 문건철 두 벌을 보관한다. 동 위원회는 그 사업진행에 필요한 기타의 보고 기록 등의 문건철 두 벌을 보관할 권한을 가진다. 동 위원회의 최후 해산시에는 상기 문건철을 쌍방에 각 한 벌씩 나누어준다.

49. 중립국감독위원회는 군사정전위원회에 본 정전협정의 수정 또는 증보에 대한 건의를 제출할 수 있다. 이러한 개정 건의는 일반적으로 더 유효한 정전을 보장할 것을 목적으로 하는 것이어야 한다.

- 남측은 현재까지 중립국감독위원회를 운영하고 있으나, 북측은 1993년도 체코슬로바키아가 체코와 슬로바키아로 분리 독립하면서 체코 대표단의 입국 자체를 거부하여 철수시켰고 1995년 폴란드 대표단을 강제 철수시켰습니다. 이후 폴란드 대표단은 유엔군측의 중립국감독위원회의 분기회의에 참석 중에 있습니다.
- 각방이 수하고 있는 : 남측과 북측이 수용(보관, 구류)하고 있는

50. 중립국감독위원회 또는 동 위원회의 매개 위원은 군사정전위원회의 임의의 위원과 통신 연락을 취할 권한을 가진다.

제3조
전쟁포로에 관한 조치

51. 본 정전협정이 효력이 발생하는 당시에 각방이 수용하고 있는 전체 전쟁포로의 석방과 송환은 본 정전협정 조인 전에 쌍방이 합의한 하기 규정에 따라 집행한다.

Armistice Agreement.

a. Within sixty(60) days after this Armistice Agreement becomes effective, each side shall, without offering any hindrance, directly repatriate and hand over in groups all those prisoners of war in its custody who insist on repatriation to the side to which they belonged at the time of capture. Repatriation shall be accomplished in accordance with the related provisions of this Article. In order to expedite the repatriation process of such personnel, each side shall, prior to the signing of the Armistice Agreement, exchange the total numbers, by nationalities, of personnel to be directly repatriated. Each group of prisoners of war delivered to the other side shall be accompanied by rosters, prepared by nationality, to include name, rank(if any) and internment or military serial number.

b. Each side shall release all those remaining prisoners of war, who are not directly repatriated, from its military control and from its custody and hand them over to the Neutral Nations Repatriation Commission for disposition in accordance with the provisions in the Annex hereto: "Term of Reference for Neutral Nations Repatriation Commission."

c. So that there may be no misunderstanding owing to the equal use of three languages, the act of delivery of a prisoner of war by one side to the other side shall, for the purposes of this Armistice Agreement, be called "repatriation" in English, "송환" (SONG HWAN) in Korean, and "遣還" (CH' IEN FAN) in Chinese, notwithstanding the nationality or place of residence of such prisoner of war.

52. Each side insures that it will not employ in acts of war in the Korean conflict any prisoner of war released and repatriated

ㄱ) 본 정전협정이 효력을 발생한 후 육십(60) 일 이내에 각방은 그 수하에 있는 송환을 견지하는 전체 전쟁포로를 포로 된 당시에 그들이 속한 일방에 집단적으로 나누어 직접 송환 인도하며 어떠한 저애도 가하지 못한다. 송환은 본 조의 각항 관계 규정에 의하여 완수한다. 이러한 인원의 송환 수속을 촉진시키기 위하여 각방은 정전협정 조인 전에 직접 송환될 인원의 국적별로 분류한 총수를 교환한다. 상대방에 인도되는 전쟁포로의 각 집단은 국적별로 작성한 명부를 휴대하되 이에는 성명, 계급(계급이 있으면) 및 수용번호 또는 군번호를 포함한다.

ㄴ) 각방은 직접 송환하지 않은 나머지 전쟁포로를 그 군사통제와 수하로부터 석방하여 두 중립국송환위원회에 넘겨 본 정전협정 부록 '중립국 송환위원회 직권의 범위' 의 각조의 규정에 의하여 처리케 한다.

ㄷ) 세 가지 글을 병용함으로 인하여 발생할 수 있는 오해를 피하기 위하여 본 정전협정의 용어로써 일방이 전쟁포로를 상대방에 인도하는 행동을 그 전쟁포로의 국적과 거주지의 여하를 불문하고 영문중에서는 "REPATRIATION" 한국문중에서는 "송환" 중국문중에서는 "送還" 이라고 규정한다.

52. 각방은 본 정전협정의 효력 발생에 의하여 석방되며 송환되는 어떠한 전쟁 포로든지 한국 충돌중의 전쟁

- 포로를 수용하고 있는 남측과 북측은 포로들이 포로 될 당시에 속한 집단에(남측 또는 북측, 등) 집단적으로 송환한다.
  포로교환 총수(병상전쟁포로와 송환희망포로)
  유엔군측: 82,493명(북한군 75,823명, 중공군 6,670명)
  공산군측: 13,457명(한국군 8,333명, 유엔군 5,124명)
  *1951년 6월 25일 북측은 한국군과 유엔군포로 규모를 108,257명이라고 공식발표하였으나, 같은 해 12월 휴전협상 포로명단 교부시 11,559명에 불과한 명단을 제공했었습니다.
- 중립국송환위원회 : 정전협정 발효일 이후 피송환권을 행사하지 않은 포로들에게 제네바 협약의 정신으로 포로의 송환 권한을 지원하기 위한 조직

- repatriation : 송환

incident to the coming into effect of this Armistice Agreement.

53. All the sick and injured prisoners of war who insist upon repatriation shall be repatriated with priority. Insofar as possible, there shall be captured medical personnel repatriated concurrently with the sick and injured prisoners of war, so as to provide medical care and attendance en route.

54. The repatriation of all of the prisoners of war required by Subparagraph 51a hereof shall be completed within a time limit of sixty(60) days after this Armistice Agreement becomes effective. Within this time limit each side undertakes to complete the repatriation of the above-mentioned prisoners of war in its custody at the earliest practicable time.

55. PANMUNJOM is designated as the place where prisoners of war will be delivered and received by both sides. Additional place(s) of delivery and reception of prisoners of war in the Demilitarized Zone may be designated, if necessary, by the Committee for Repatriation of Prisoners of War.

56. a. A Committee for Repatriation of Prisoners of War is hereby established. It shall be composed of six(6) officers of field grade, three(3) of whom shall be appointed by the Commander-in-Chief, United Nations Command, and three(3) of whom shall be appointed jointly by the Supreme Commander of the Korean People's Army and the Commander of the Chinese People's Volunteers. This Committee shall, under the general supervision and direction of the Military Armistice Commission, be responsible for coordinating the

행동에 사용하지 않을 것을 보장한다.

53. 송환을 견지하는 전체 병상전쟁포로는 우선적으로 송환한다. 가능한 범위내에서 포함된 의무인원을 병상전쟁포로와 동시에 송환하여 도중에서 의료와 간호를 제공하도록 한다.

• 병상전쟁포로(1953. 4. 20. ~5. 3일 송환) 유엔군측 6,670명, 공산군측 684명

54. 본 정전협정 제51항 ㄱ에 규정한 전체 전쟁포로의 송환은 본 정전협정이 효력을 발생한 후 육십(60) 일의 기한내에 완료한다. 이 기한내에 각방은 책임지고 그가 수용하고 있는 상기 전쟁포로의 송환을 실제 가능한한 속히 완료한다.

55. 판문점을 쌍방의 전쟁포로 인도 인수 지점으로 정한다. 필요할 때에는 전쟁 포로송환위원회는 기타의 전쟁포로 인도 인수지점(들)을 비무장지대내에 증설할 수 있다.

56. ㄱ) 전쟁포로송환위원회를 설립한다. 동 위원회는 영관급 장교 육(6) 명으로 구성하되 그 중 삼(3) 명은 국제연합군 총사령관이 이를 임명하며 그 중 삼 (3) 명은 조선인민군 최고사령관과 중국인민지원군 사령원이 공동으로 이를 임명한다. 동 위원회는 군사정전위원회의 전반적 감독과 지도하에서 책임지고 쌍방의 전쟁포로 송환에 관계되는 구체적 계획을 조절하며 쌍방이 본 정전협정중의 전쟁포로 송환에 관계되는 일

specific plans of both sides for the repatriation of prisoners of war and for supervising the execution by both sides of all of the provisions of this Armistice Agreement relating to the repatriation of prisoners of war. It shall be the duty of this Committee to coordinate the timing of the arrival of prisoners of war at the place(s) of delivery and reception of prisoners of war from the prisoner of war camps of both sides; to make, when necessary, such special arrangements as may be required with regard to the transportation and welfare of sick and injured prisoners of war; to coordinate the work of the joint Red Cross teams, established in Paragraph 57 hereof, in assisting in the repatriation of prisoners of war; to supervise the implementation of the arrangements for the actual repatriation of prisoners of war stipulated in Paragraph 53 and 54 hereof; to select, when necessary, additional place(s) of delivery and reception of prisoners of war; to arrange for security at the place(s) of delivery and reception of prisoners of war; and to carry out such other related functions as are required for the repatriation of prisoners of war.

b. When unable to reach agreement on any matter relating to its responsibilities, the Committee for Repatriation of Prisoners of War shall immediately refer such matter to the Military Armistice Commission for decision. The Committee for Repatriation of Prisoners of War shall maintain its headquarters in proximity to the headquarters of the Military Armistice Commission.

c. The Committee for Repatriation of Prisoners of War shall be dissolved by the Military Armistice Commission upon completion of the program of repatriation of prisoners of war.

57. a. Immediately after this Armistice Agreement becomes effective, joint Red Cross teams composed of representatives of the national

체 규정을 실시하는 것을 감독한다. 동 위원회의 임무는 전쟁포로들이 쌍방 전쟁포로 수용소로부터 전쟁포로 인도 인수 지점(들)에 도달하는 시간을 조절하며 필요할 때에는 병상전쟁 포로의 수송 및 복리에 요구되는 특별한 조치를 취하며 본 정전협정 제57항에서 설립된 공동적십자소조의 전쟁포로 송환 협조사업을 조절하며 본 정전협정 제53항과 제54항에 규정한 전쟁포로 실제 송환 조치의 실시를 감독하며 필요할 때에는 추가적 전쟁포로 인도 인수 지점(들)을 선정하며 전쟁포로의 인도 인수 지점(들)의 안전 조치를 취하며 전쟁포로 송환에 필요한 기타 관계 임무를 집행하는 것이다.

ㄴ) 전쟁포로송환위원회는 그 임무에 관계되는 어떠한 사항에 대하여 합의에 도달하지 못할 때에는 이러한 사항을 즉시로 군사정전위원회에 제기하여 결정하도록 한다. 전쟁포로송환위원회는 군사정전위원회 본부 부근에 그 본부를 설치한다.

ㄷ) 전쟁포로송환위원회가 전쟁포로 송환계획을 완수한 때에는 군사정전위원회가 즉시로 이를 해산시킨다.

• 임무종료 후 해산 : 1954년

57. ㄱ) 본 정전협정이 효력을 발생한 후 즉시로 국제연합군에 군대를 제공하고 있는 각국의 적십자사 대표

Red Cross Societies of the countries contributing forces to the United Nations Command on the one hand, and representatives of the Red Cross Society of the Democratic People's Republic of Korea and representatives of the Red Cross Society of the People's Republic of China on the other hand, shall be established. The joint Red Cross teams shall assist in the execution by both sides of those provisions of this Armistice Agreement relating to the repatriation of all the prisoners of war specified in Sub-paragraph 51a hereof, who insist upon repatriation, by the performance of such humanitarian services as are necessary and desirable for the welfare of the prisoners of war. To accomplish this task, the joint Red Cross teams shall provide assistance in the delivering and receiving of prisoners of war by both sides at the place(s) of delivery and reception of prisoners of war, and shall visit the prisoner of war camps of both sides to comfort the prisoners of war and to bring in and distribute gift articles for the comfort and welfare of the prisoners of war. The joint Red Cross teams may provide services to prisoners of war while en route from prisoner of war camps to the place(s) of delivery and reception of prisoners of war.

b. The joint Red Cross teams shall be organized as set forth below:

(1) One team shall be composed of twenty(20) members, namely, ten(10) representatives from the national Red Cross Societies of each side, to assist in the delivering and receiving of prisoners of war by both sides at the place(s) of delivery and reception of prisoners of war. The chairmanship of this team shall alternate daily between representatives from the Red Cross Societies of the two sides. The work and services of this team shall be coordinated by the Committee for Repatriation of Prisoners of War.

를 일방으로 하고 조선민주주의인민공화국 적십자사 대표와 중화인민공화국 적십자사 대표를 다른 일방으로 하여 조직되는 공동적십자소조를 설립한다. 공동적십자소조는 전쟁포로의 복리에 요망되는 인도주의적 복무로써 쌍방이 본 정전협정 제51항 ㄱ에 규정한 송환을 견지하는 전체 전쟁포로의 송환에 관계되는 규정을 집행하는 것을 협조한다. 이 임무를 완수하기 위하여 공동적십자소조는 전쟁포로 인도인수지점(들)에서 쌍방의 전쟁포로 인도인수 사업을 협조하며 쌍방의 전쟁포로 수용소를 방문하여 위문하며 전쟁포로의 위문과 전쟁포로의 복리를 위한 선물을 가지고 가서 분배한다. 공동적십자소조는 전쟁포로수용소에서 전쟁포로 인도인수지점(들)으로 가는 도중에 있는 전쟁포로에게 복무를 제공할 수 있다.

ㄴ) 공동적십자소조는 다음과 같은 규정에 의하여 조직한다.

(1) 한 소조는 각방의 본국 적십자사로부터 각기 대표 십(10) 명씩을 내어 쌍방 합하여 이십(20) 명으로 구성하며 전쟁포로 인도인수지점(들)에서 쌍방의 전쟁포로의 인도인수를 협조한다. 동 소조의 의장은 쌍방 적십자사 대표가 매일 윤번으로 담당한다. 동 소조의 사업과 복무는 전쟁포로송환위원회가 이를 조절한다.

(2) One team shall be composed of sixty(60) members, namely, thirty(30) representatives from the national Red Cross Societies of each side, to visit the prisoner of war camps under the administration of the Korean People's Army and the Chinese People's Volunteers. This team may provide services to prisoners of war while en route from the prisoner of war camps to the place(s) of delivery and reception of prisoners of war. A representative of the Red Cross Society of the Democratic People's Republic of Korea or of the Red Cross Society of the People's Republic of China shall serve as chairman of this team.

(3) One team shall be composed of sixty(60) members, namely, thirty(30) representatives from the national Red Cross Societies of each side, to visit the prisoner of war camps under the administration of the United Nations Command. This team may provide services to prisoners of war while en route from the prisoner of war camps to the place(s) of delivery and reception of prisoners of war. A representative of a Red Cross Society of a nation contributing forces to the United Nations Command shall serve as chairman of this team.

(4) In order to facilitate the functioning of each joint Red Cross team, sub-teams composed of not less than two(2) members from the teams, with an equal number of representatives from each side, may be formed as circumstances require.

(5) Additional personnel such as drivers, clerks, and interpreters, and such equipment as may be required by the joint Red Cross teams to perform their missions, shall be furnished by the Commander of each side to the team operating in the territory under his military control.

(6) Whenever jointly agreed upon by the representatives of both

(2) 한 소조는 각방의 본국 적십자사로부터 각기 대표 삼십(30) 명씩을 내어 쌍방 합하여 육십(60) 명으로 구성하며 조선인민군 및 중국인민지원군 관리하의 전쟁포로 수용소를 방문하며 또 전쟁포로수용소에서 전쟁포로 인도인수지점(들)으로 가는 도중에 있는 전쟁포로에게 복무를 제공할 수 있다. 조선민주주의인민공화국 적십자사 또는 중화인민공화국 적십자사의 대표가 동 소조의 의장을 담당한다.

(3) 한 소조는 각방의 본국 적십자사로부터 각기 대표 삼십(30) 명씩을 내어 쌍방 합하여 육십(60) 명으로 구성하며 국제연합군 관리하의 전쟁포로수용소를 방문하며 또 전쟁포로수용소에서 전쟁포로 인도인수지점(들)으로 가는 도중에 있는 전쟁포로에게 복무를 제공할 수 있다. 국제연합군에 군대를 제공하고 있는 한 나라의 적십자사 대표가 동 소조의 의장을 담당한다.

(4) 각 공동적십자소조의 임무 집행의 편의를 위하여 정황이 필요로 할 때에는 최소 이(2) 명의 소조원으로 구성하는 분조를 설립할 수 있다. 분조내에서 각방은 동등한 수의 대표를 가진다.
(5) 각방 사령관은 그의 군사통제지역내에서 사업하는 공동적십자소조에 운전수, 서기 및 통역과 같은 부속인원 및 각 소조가 그 임무 집행상 필요로 하는 장비를 공급한다.

(6) 어떠한 공동적십자소조든지 동 소조의 쌍방 대표

sides on any joint Red Cross team, the size of such team may be increased or decreased, subject to confirmation by the Committee for Repatriation of Prisoners of War.

c. The Commander of each side shall cooperate fully with the joint Red Cross teams in the performance of their functions, and undertakes to insure the security of the personnel of the joint Red Cross team in the area under his military control. The Commander of each side shall provide such logistic, administrative, and communications facilities as may be required by the team operating in the territory under his military control.

d. The joint Red Cross teams shall be dissolved upon completion of the program of repatriation of all the prisoners of war specified in Sub-paragraph 51(a) hereof, who insist upon repatriation.

58. a. The Commander of each side shall furnish to the Commander of the other side as soon as practicable, but not later than ten(10) days after this Armistice Agreement becomes effective, the following information concerning prisoners of war:

(1) Complete data pertaining to the prisoners of war who escaped since the effective date of the data last exchanged

(2) Insofar as practicable, information regarding name, nationality, rank, and other identification data, date and cause of death, and place of burial, of those prisoners of war who died while in his custody.

b. If any prisoners of war escape or die after the effective date of the supplementary information specified above, the detaining side shall furnish to the other side, through the Committee for Repatriation of Prisoners of War, the data pertaining thereto in accordance with the provisions of Sub-paragraph 58(a) hereof. Such data shall be

가 동의하는 때에는 그 인원수를 증감할 수 있다. 단 이는 전쟁포로송환위원회의 인가를 거쳐야 한다.

ㄷ) 각방 사령관은 공동적십자소조가 그의 임무를 집행하는데 충분한 협조를 주며 또 그의 군사통제지역내에서 책임지고 공동적십자소조 인원들의 안전을 보장한다. 각방 사령관은 그의 군사통제지역내에서 사업하는 이러한 소조에 요구되는 보급, 행정 및 통신상의 편의를 준다.

ㄹ) 공동적십자소조는 본 정전협정 제51항 ㄱ에 규정한 송환을 견지하는 전체 전쟁포로의 송환 계획이 완수되었을 때에는 즉시로 해산한다.

58. ㄱ) 각방 사령관은 가능한 범위내에서 속히 그러나 본 정전협정이 효력을 발생한 후 십(10) 일 이내에 상대방 사령관에게 다음과 같은 전쟁포로에 관한 재료를 제공한다.

(1) 제일 마지막 번에 교환한 자료의 마감한 날짜 이후에 도망한 전쟁포로에 관한 완전한 자료.

(2) 실제로 실행할 수 있는 범위내에서 수용기간 중에 사망한 전쟁포로의 성명, 국적, 계급별 및 기타의 식별자료 또한 사망날짜, 사망원인 및 매장 지점에 관한 재료.

ㄴ) 만일 위에 규정한 보충 재료의 마감한 날짜 이후에 도망하였거나 또는 사망한 어떠한 전쟁포로가 있으면 수용한 일방은 본 조 제58항 ㄱ의 규정에 의하여 관계 자료를 전쟁포로송환위원회를 거쳐 상대방에 제공한다. 이러한 자료는 전쟁포로 인도인수 계획을 완

furnished at ten-day intervals until the completion of the program of delivery and reception of prisoners of war.

c. Any escaped prisoner of war who returns to the custody of the detaining side after the completion of the program of delivery and reception of prisoners of war shall be delivered to the Military Armistice Commission for disposition.

59. a. All civilians who, at the time this Armistice Agreement becomes effective, are in territory under the military control of the Commander-in-Chief, United Nations Command, and who, on 24 June 1950, resided north of the Military Demarcation Line established in this Armistice Agreement shall, if they desire to return home, be permitted and assisted by the Commander-in-Chief, United Nations Command, to return to the area north of the Military Demarcation Line; and all civilians who, at the time this Armistice Agreement becomes effective, are in territory under the military control of the Supreme Commander of the Korean People's Army and the Commander of the Chinese People's Volunteers, and who, on 24 June 1950, resided south of the Military Demarcation Line established in this Armistice Agreement shall, if they desire to return home, be permitted and assisted by the Supreme Commander of the Korean People's Army and the Commander of the Chinese People's Volunteers to return to the area south of the Military Demarcation Line. The Commander of each side shall be responsible for publicizing widely throughout territory under his military control the contents of the provisions of this Sub-paragraph, and for calling upon the appropriate civil authorities to give necessary guidance and assistance to all such civilians who desire to return home.

b. All civilians of foreign nationality who, at the time this Armistice

수할 때까지 십(10) 일에 일차씩 제공한다.

ㄷ) 전쟁포로 인도인수 계획을 완수한 후에 본래 수용하고 있던 일방에 다시 돌아온 도망하였던 어떠한 전쟁포로도 이를 군사정전위원회에 넘기어 처리한다.

59. ㄱ) 본 정전협정이 효력을 발생하는 당시에 국제연합군 총사령관의 군사통제지역에 있는 자로서 1950년 6월 24일에 본 정전협정에 확정된 군사분계선 이북에 거주한 전체 사민에 대하여서는 그들이 귀향하기를 원한다면 국제연합군 총사령관은 그들이 군사분계선 이북지역에 돌아가는 것을 허용하며 협조한다. 본 정전협정이 효력을 발생하는 당시에 조선인민군 최고사령관과 중국인민지원군 사령원의 군사통제지역에 있는 자로서 1950년 6월 24일에 본 정전협정에 확정된 군사분계선 이남에 거주한 전체 사민에 대하여서는 그들이 귀향하기를 원한다면 조선인민군 최고사령관과 중국인민지원군 사령원은 그들이 군사분계선 이남지역에 돌아가는 것을 허용하며 협조한다. 각방 사령관은 책임지고 본 규정의 내용을 그의 군사통제지역에 광범히 선포하며 또 적당한 민정 당국을 시켜 귀향하기를 원하는 이러한 전체 사민에게 필요한 지도와 협조를 주도록 한다.

ㄴ) 본 정전협정이 효력을 발생하는 당시에 조선인민

Agreement becomes effective, are in territory under the military control of the Supreme Commander of the Korean People's Army and the Commander of the Chinese People's Volunteers shall, if they desire to proceed to territory under the military control of the Commander-in-Chief, United Nations Command, be permitted and assisted to do so; all civilians of foreign nationality who, at the time this Armistice Agreement becomes effective, are in territory under the military control of the Commander-in-Chief, United Nations Command, shall, if they desire to proceed to territory under the military control of the Supreme Commander of the Korean People's Army and the Commander of the Chinese People's Volunteers, be permitted and assisted to do so. The Commander of each side shall be responsible for publicizing widely throughout the territory under his military control the contents of the provisions of this Sub-paragraph, and for calling upon the appropriate civil authorities to give necessary guidance and assistance to all such civilians of foreign nationality who desire to proceed to territory under the military control of the Commander of the other side.

c. Measures to assist in the return of civilians provided for in Sub-paragraph 59(a) hereof and the movement of civilians provided for in Sub-paragraph 59(b) hereof shall be commenced by both sides as soon as possible after this Armistice Agreement becomes effective.

d. (1) A Committee for Assisting the Return of Displaced Civilians is hereby established. It shall be composed of four(4) officers of field grade, two(2) of whom shall be appointed by the Commander-in-Chief, United Nations Command, and two(2) of whom shall be appointed jointly by the Supreme Commander of the Korean People's Army and the Commander of the Chinese People's Volunteers. This Committee shall, under the general supervision and

군 최고사령관과 중국인민지원군 사령원의 군사통제지역에 있는 전체 외국적의 사민중 국제연합군 총사령관의 군사통제지역으로 가기를 원하는 자에게는 그가 국제연합군 총사령관의 군사통제지역으로 가는 것을 허용하며 협조한다. 본 정전협정이 효력을 발생하는 당시에 국제연합군 총사령관의 군사통제지역에 있는 전체 외국적의 사민중 조선인민군 최고사령관과 중국인민지원군 사령원의 군사통제지역으로 가기를 원하는 자에게는 그가 조선인민군 최고사령관과 중국인민지원군 사령원의 군사통제지역으로 가는 것을 허용하며 협조한다. 각방 사령관은 책임지고 본 규정의 내용을 그의 군사통제지역에 광범히 선포하며 또 적당한 민정 당국을 시켜 상대방 사령관의 군사통제지역으로 가기를 원하는 이러한 전체 외국적의 사민에게 필요한 지도와 협조를 주도록 한다.

ㄷ) 쌍방이 본 조 제59항 ㄱ에 규정한 사민의 귀향과 본 조 제59항 ㄴ에 규정한 사민의 이동을 협조하는 조치는 본 정전협정이 효력을 발생한 후 될 수 있는 한 속히 개시한다.
ㄹ) (1) 실향사민귀향협조위원회를 설립한다.
동 위원회는 영급 장교 사(4) 명으로 구성하되 그중 이(2) 명은 국제연합군 총사령관이 이를 임명하며 그중 이(2) 명은 조선인민군 최고사령관과 중국인민지원군 사령원이 공동으로 이를 임명한다. 동 위원회는 군사정전위원회의 전반적 감독과 지도 밑에 책임지고 상기 사민의 귀향을 협조하는데 관계되는 쌍방의 구체

direction of the Military Armistice Commission, be responsible for coordinating the specific plans of both sides for assistance to the return of the above-mentioned civilians, and for supervising the execution by both sides of all of the provisions of this Armistice Agreement relating to the return of the above-mentioned civilians. It shall be the duty of this Committee to make necessary arrangements, including those of transportation, for expediting and coordinating the movement of the above-mentioned civilians; to select the crossing point(s) through which the above-mentioned civilians will cross the Military Demarcation Line; to arrange for security at the crossing point(s); and to carry out such other functions as are required to accomplish the return of the above-mentioned civilians.

(2) When unable to reach agreement on any matter relating to its responsibilities, the Committee for Assisting the Return of Displaced Civilians shall immediately refer such matter to the Military Armistice Commission for decision. The Committee for Assisting the Return of Displaced Civilians shall maintain its headquarters in proximity to the headquarters of the Military Armistice Commission.

(3) The Committee for Assisting the Return of Displaced Civilians shall be dissolved by the Military Armistice Commission upon fulfillment of its mission.

적 계획을 조절하며 또 상기 사민의 귀향에 관계되는 본 정전협정중의 일체 규정을 쌍방이 집행하는 것을 감독한다. 동 위원회의 임무는 운수 조치를 포함한 필요한 조치를 취함으로써 상기 사민의 이동을 촉진 및 조절하며 상기 사민이 군사분계선을 통과하는 월경지점(들)을 선정하며 월경지점(들)의 안전조치를 취하며 또 상기 사민 귀향을 완료하기 위하여 필요한 기타 임무를 집행하는 것이다.

(2) 실향사민귀향협조위원회는 그의 임무에 관계되는 어떠한 사항이든지 합의에 도달할 수 없을 때에는 이를 곧 군사정전위원회에 제출하여 결정하게 한다. 실향사민귀향협조위원회는 그의 본부를 군사정전위원회의 본부 부근에 설치한다.

(3) 실향사민귀향협조위원회가 그의 임무를 완수한 때에는 군사정전위원회가 즉시로 이를 해산시킨다.

ARTICLE IV
RECOMMENDATION TO THE GOVERNMENTS CONCERNED ON BOTH SIDES

60. In order to insure the peaceful settlement of the Korean question, the military Commanders of both sides hereby recommend to the governments of the countries concerned on both sides that, within three(3) months after the Armistice Agreement is signed and becomes effective, a political conference of a higher level of both sides be held by representatives appointed respectively to settle through negotiation the questions of the withdrawal of all foreign forces from Korea, the peaceful settlement of the Korean question, etc.

ARTICLE V
MISCELLANEOUS

61. Amendments and additions to this Armistice Agreement must be mutually agreed to by the Commanders of the opposing sides.

62. The Articles and Paragraphs of this Armistice Agreement shall remain in effect until expressly superseded either by mutually acceptable amendments and additions or by provision in an appropriate agreement for a peaceful settlement at a political level between both sides.

63. All of the provisions of this Armistice Agreement, other than Paragraph 12, shall become effective at 2200 hours on 27 July 1953.

제4조
쌍방 관계정부들에의 건의

60. 한국 문제의 평화적 해결을 보장하기 위하여 쌍방 군사사령관은 쌍방의 관계 각국 정부에 정전협정이 조인되고 효력을 발생한 후 삼(3) 개월내에 각기 대표를 파견하여 쌍방의 한 급 높은 정치회의를 소집하고 한국으로부터의 모든 외국군대의 철거 및 한국 문제의 평화적 해결 등 문제들을 협의할 것을 이에 건의한다.

• 정치회의는 1954년 2월 28일 미 · 영 · 불 · 소 4개국 외무부장관간의 베를린회담, 1954년 4월 26일~6월 15일까지 19개국 외무부장관간의 제네바회담에서 실시되었으나 유엔군측은 한반도 문제는 유엔으로 이관할 수밖에 없다는 요지의 공동선언을 발표하고 성과 없이 종료되었습니다.(한국전쟁 참전국 15개국(남아공 불참)과 남 · 북한, 중국, 소련 참석)

제5조
부칙

61. 본 정전협정에 대한 수정과 증보는 반드시 적대쌍방 사령관들의 호상 합의를 거쳐야 한다.

• 적대 쌍방은 공산군측과 유엔군측입니다.

62. 본 정전협정의 각 조항은 쌍방이 공동으로 접수하는 수정 및 증보 또는 쌍방의 정치적 수준에서의 평화적 해결을 위한 적당한 협정 중의 규정에 의하여 명확히 교체될 때까지는 계속 효력을 가진다.

63. 제12항을 제외한 본 정전협정의 일체 규정은 1953년 7월 27일 2200시부터 효력을 발생한다.

Done at Panmunjom, Korea, at 1000 hours on the 27th day of July, 1953, in English, Korean, and Chinese, all text being equally authentic.

KIM IL SUNG
Marshal, Democratic People's Republic of Korea
Supreme Commander, Korean People's Army

PENG TEH-HUAI
Commander, Chinese People's Volunteers

MARK W. CLARK
General, United States Army
Commander-in-Chief, United Nations Command

PRESENT

NAM IL
General, Korean People's Army
Senior Delegate,
Delegation of the Korean People's Army and the Chinese People's Volunteers

WILLIAM K. HARRISON, JR.
Lieutenant General, United States Army
Senior Delegate,
United Nations Command Delegation

1953년 7월 27일 1000시에 한국 판문점에서 영문, 한국문 및 중국문으로써 작성한다. 이 세 가지 글의 각 협정 문본은 동등한 효력을 가진다.

조선인민군 최고사령관
조선민주주의인민공화국원수
김일성

중국인민지원군
사령원
팽덕회

국제연합군 총사령관
미국 륙군 대장
마크 더불유.클라크

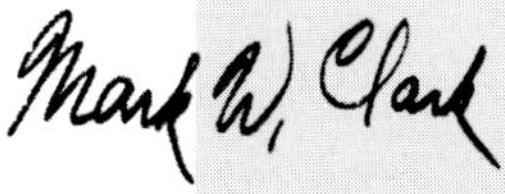

참석자

조선인민군 및
중국인민지원군 대표단
수석 대표
조선인민군 대장
남일

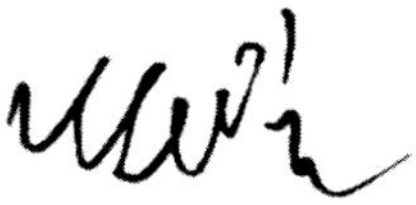

국제연합군 대표단
수석 대표
미국 륙군 중장

윌리암 케이.해리슨

ARMISTICE AGREEMENT
VOLUME II
MAPS

정 전 협 정
제 2 권
지 도

停 戰 協 定
卷 二
地圖

MAPS AUTHENTICATED:

지도를 인증함

J. C. Murray
Col USMC

For the Supreme Commander of the Korean People's Army and the Commander of the Chinese People's Volunteers

조선인민군 최고사령관과
중국인민지원군 사령원
대리

朝鮮人民軍最高司令官
與
中國人民志願軍司令員
代理簽證附圖者

For the Commander-in-Chief, United Nations Command

국제련합군 총사령관
대리

聯合國軍總司令
代理簽證附圖者

INDEX TO MAPS

지도목록

첨부한 지도 제1도의 1부터 9까지	군사분계선 및 비무장지대의 북경계선과 남경계선
첨부한 지도 제2도	군사정전위원회 감독하의 한강하구 구역
첨부한 지도 제3도	한국 서부 연해섬들의 통제
첨부한 지도 제4도	주요 교통선
첨부한 지도 제5도의1	신의주
첨부한 지도 제5도의2	청진
첨부한 지도 제5도의3	흥남
첨부한 지도 제5도의4	만포
첨부한 지도 제5도의5	신안주
첨부한 지도 제5도의6	인천
첨부한 지도 제5도의7	대구
첨부한 지도 제5도의8	부산
첨부한 지도 제5도의9	강릉
첨부한 지도 제5도의10	군산

첨부한 지도 제1도의1
군사 분계선 및
비무장 지대의 북 경계선과 남 경계선
(제2항 및 제3항을 보라)

MAP 1-1
THE MILITARY DEMARCATION LINE AND THE NORTHERN AND SOUTHERN BOUNDARIES OF THE DEMILITARIZED ZONE
(See Paragraphs 2 and 3)

附圖一之一
軍事分界線及非軍事區的北線與
（見第二款與第三款）

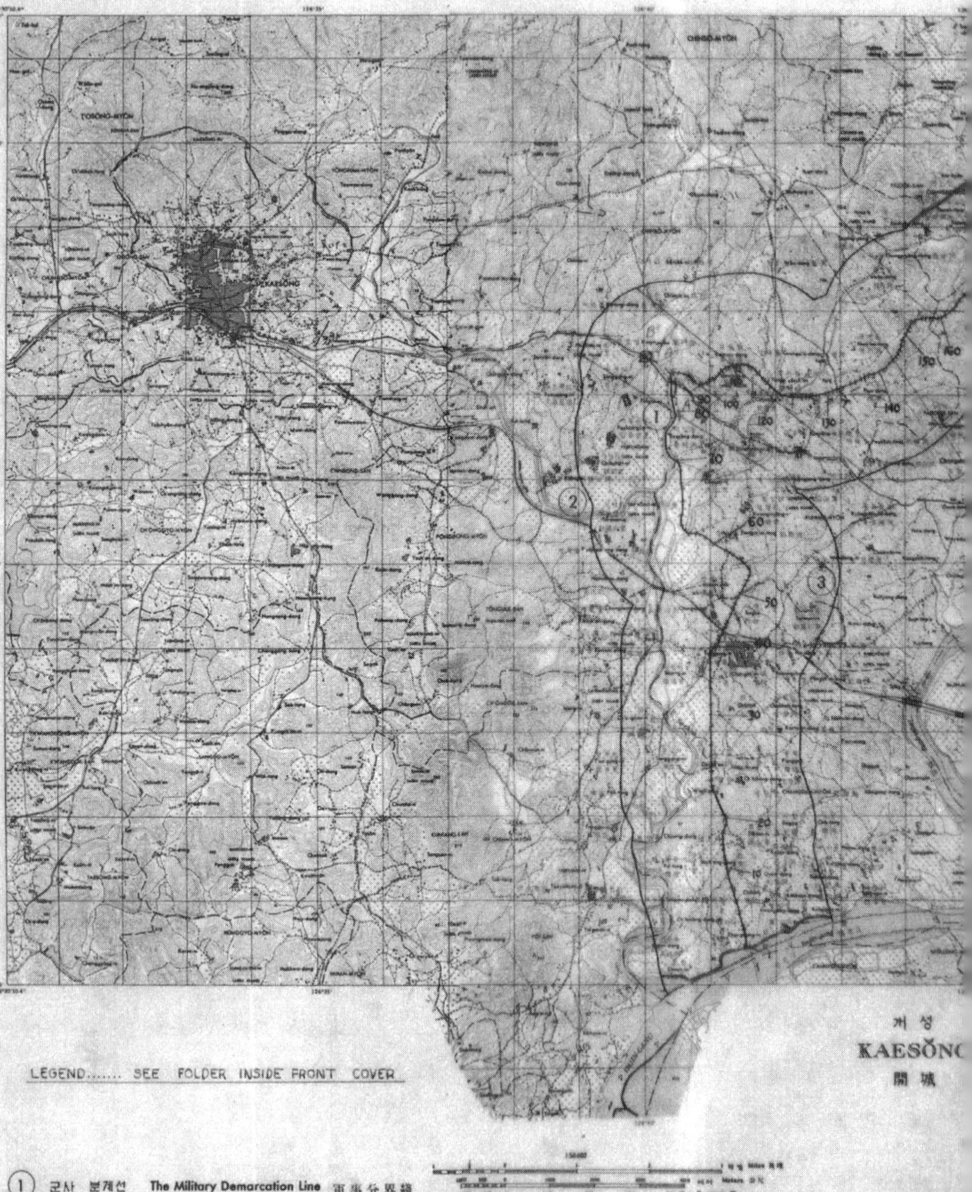

LEGEND....... SEE FOLDER INSIDE FRONT COVER

① 군사 분계선 The Military Demarcation Line 軍事分界線

② 비무장 지대의 북 경계선 The Northern Boundary of the Demilitarized Zone 非軍事區的北線

③ 비무장 지대의 남 경계선 The Southern Boundary of the Demilitarized Zone 非軍事區的南線

AUTHENTICATED

1 APRIL [illegible]
DATE

MAJ ...
STAFF OFF...

COL. ...
STAFF OFF...

첨부한 지도 제1도의2
군사 분계선 및
비무장 지대의 북 경계선과 남 경계선
(제2항 및 제3항을 보라)

MAP 1-2
THE MILITARY DEMARCATION LINE AND THE NORTHERN AND SOUTHERN BOUNDARIES OF THE DEMILITARIZED ZONE
(See Paragraphs 2 and 3)

附圖一之二
軍事分界線及非軍事區的北緣與南緣
(見第二款與第三款)

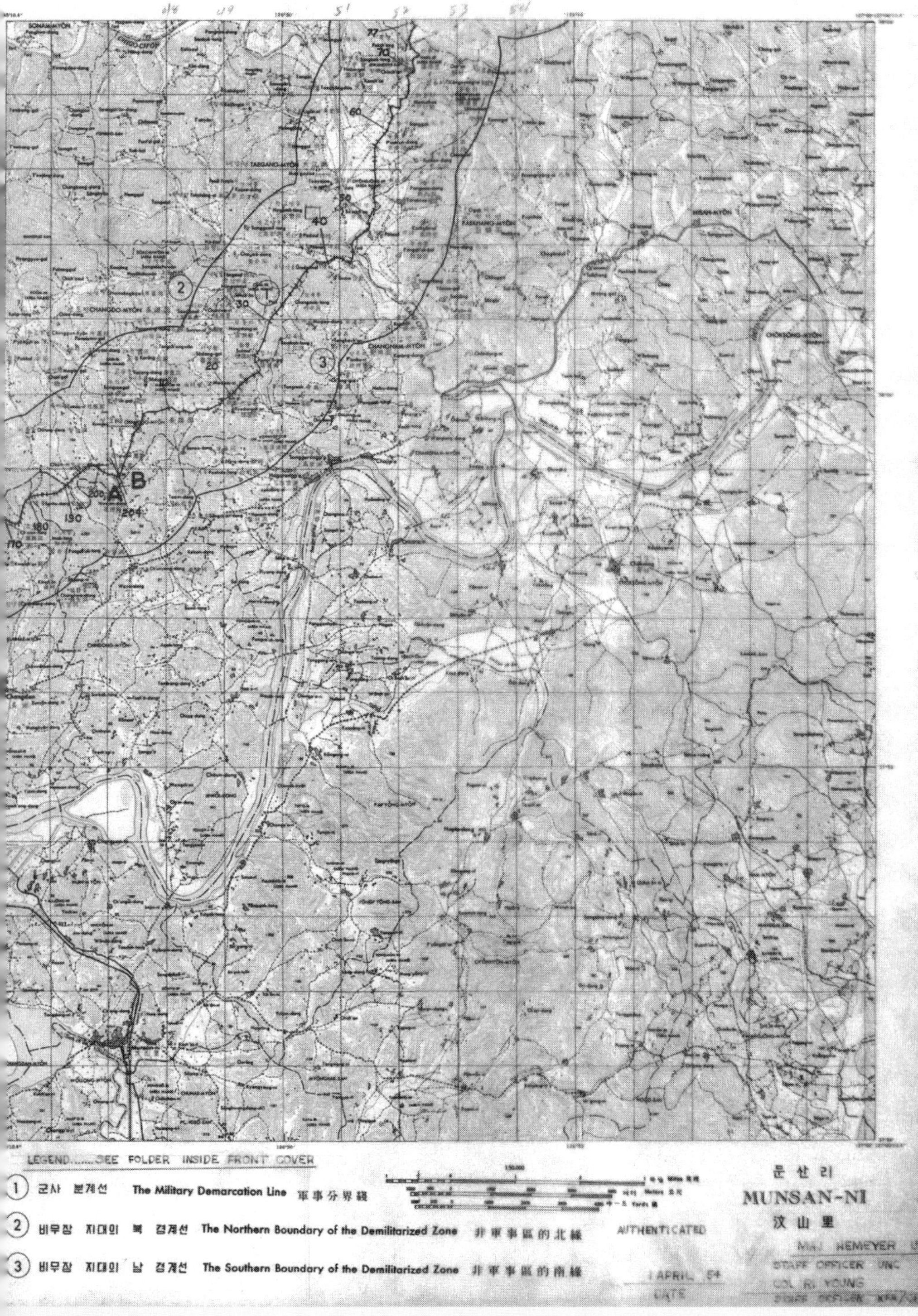

LEGEND......SEE FOLDER INSIDE FRONT COVER

1) 군사 분계선 The Military Demarcation Line 軍事分界線

2) 비무장 지대의 북 경계선 The Northern Boundary of the Demilitarized Zone 非軍事區的北緣

3) 비무장 지대의 남 경계선 The Southern Boundary of the Demilitarized Zone 非軍事區的南緣

문산리
MUNSAN-NI
汶山里

AUTHENTICATED

1 APRIL 54
DATE

MAJ HEMEYER USA
STAFF OFFICER UNC
COL RI YOUNG
[illegible] KPA/[illegible]

첨부한 지도 제1도의3
군사 분계선 및
비무장 지대의 북 경계선과 남 경계선
(제2항 및 제3항을 보라)

MAP 1-3
THE MILITARY DEMARCATION LINE AND THE NORTHERN AND SOUTHERN BOUNDARIES OF THE DEMILITARIZED ZONE
(See Paragraphs 2 and 3)

附圖一之三
軍事分界線及非軍事區的北緣與南
(見第二款與第三款)

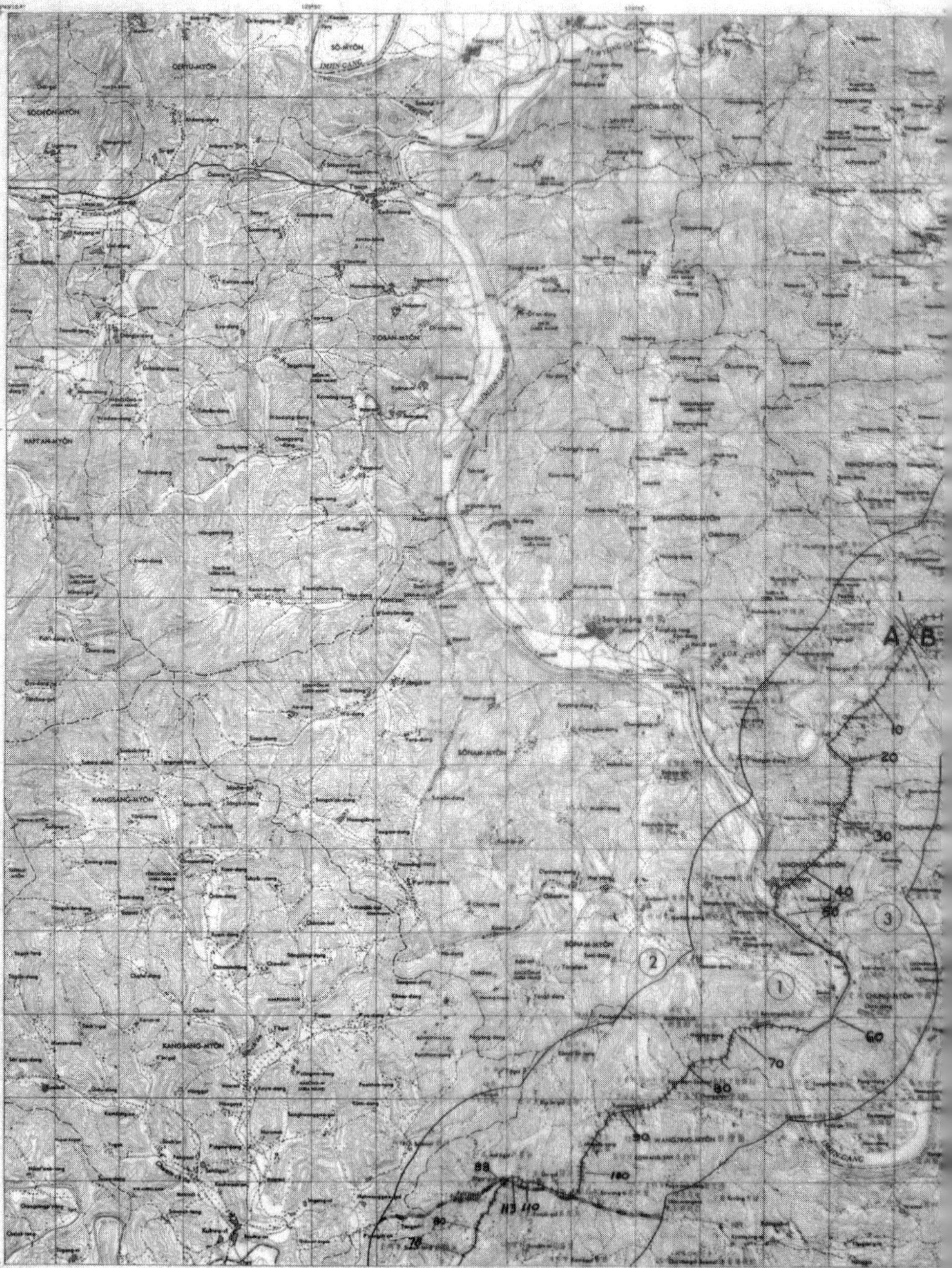

LEGEND......SEE FOLDER INSIDE FRONT COVER

① 군사 분계선 The Military Demarcation Line 軍事分界線

② 비무장 지대의 북 경계선 The Northern Boundary of the Demilitarized Zone 非軍事區的北緣

③ 비무장 지대의 남 경계선 The Southern Boundary of the Demilitarized Zone 非軍事區的南緣

AUTHENTICATED

APRIL 54
DATE

삭령
SANGNYŎ
朔寧
STAFF OFFIC

첨부한 지도 제1도의 4
군사 분계선 및
비무장 지대의 북 경계선과 남 경계선
(제2항 및 제3항을 보라)

MAP 1-4
THE MILITARY DEMARCATION LINE AND THE NORTHERN AND SOUTHERN BOUNDARIES OF THE DEMILITARIZED ZONE
(See Paragraphs 2 and 3)

附圖一之四
軍事分界線及非軍事區的北線與南線
(見第二款與第三款)

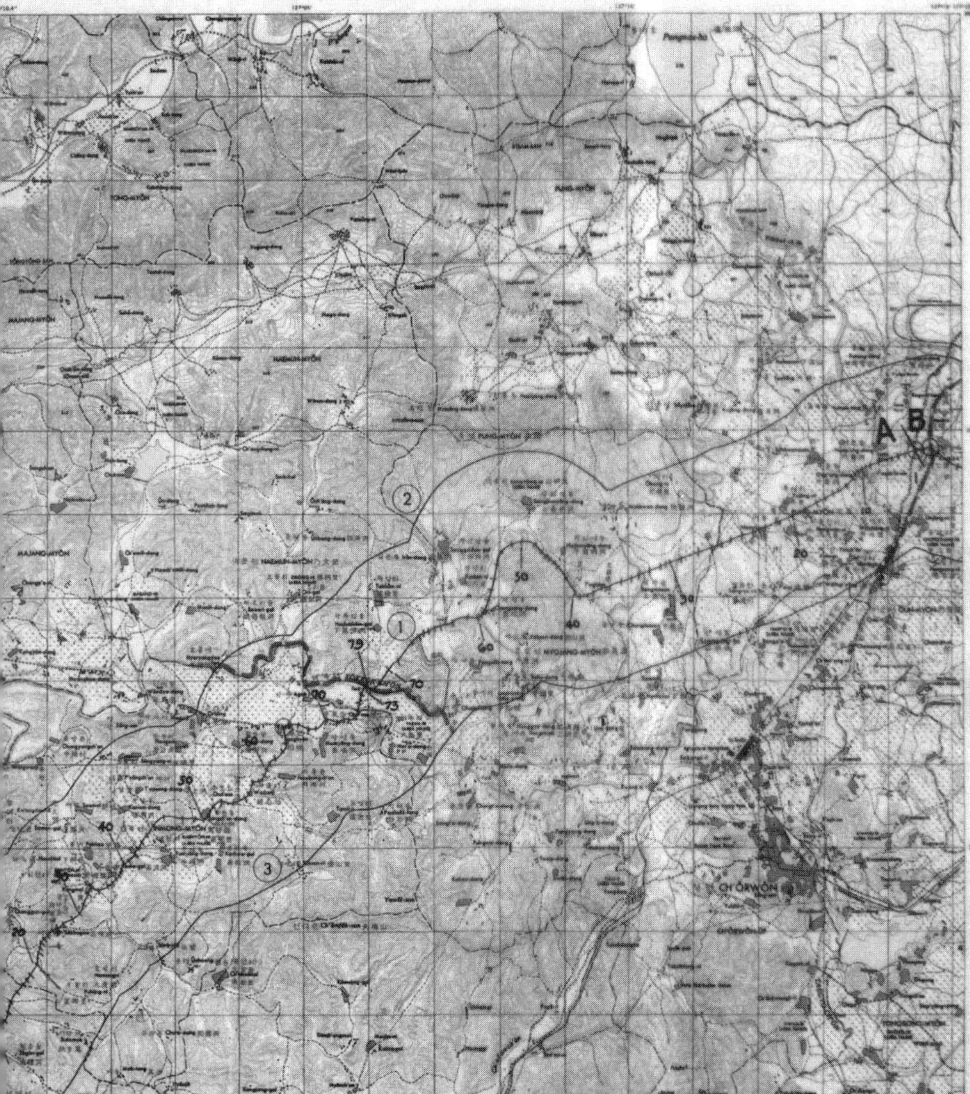

1) 군사 분계선 The Military Demarcation Line 軍事分界線

2) 비무장 지대의 북 경계선 The Northern Boundary of the Demilitarized Zone 非軍事區的北線

3) 비무장 지대의 남 경계선 The Southern Boundary of the Demilitarized Zone 非軍事區的南線

철원
CH'ORWON
鐵原

첨부한 지도 제1도의 5
군사 분계선 및
비무장 지대의 북 경계선과 남 경계선
(제2항 및 제3항을 보라)

MAP 1-5
THE MILITARY DEMARCATION LINE AND THE NORTHERN AND SOUTHERN BOUNDARIES OF THE DEMILITARIZED ZONE
(See Paragraphs 2 and 3)

附圖一之五
軍事分界線及非軍事區的北線與南
(見第二款與第三款)

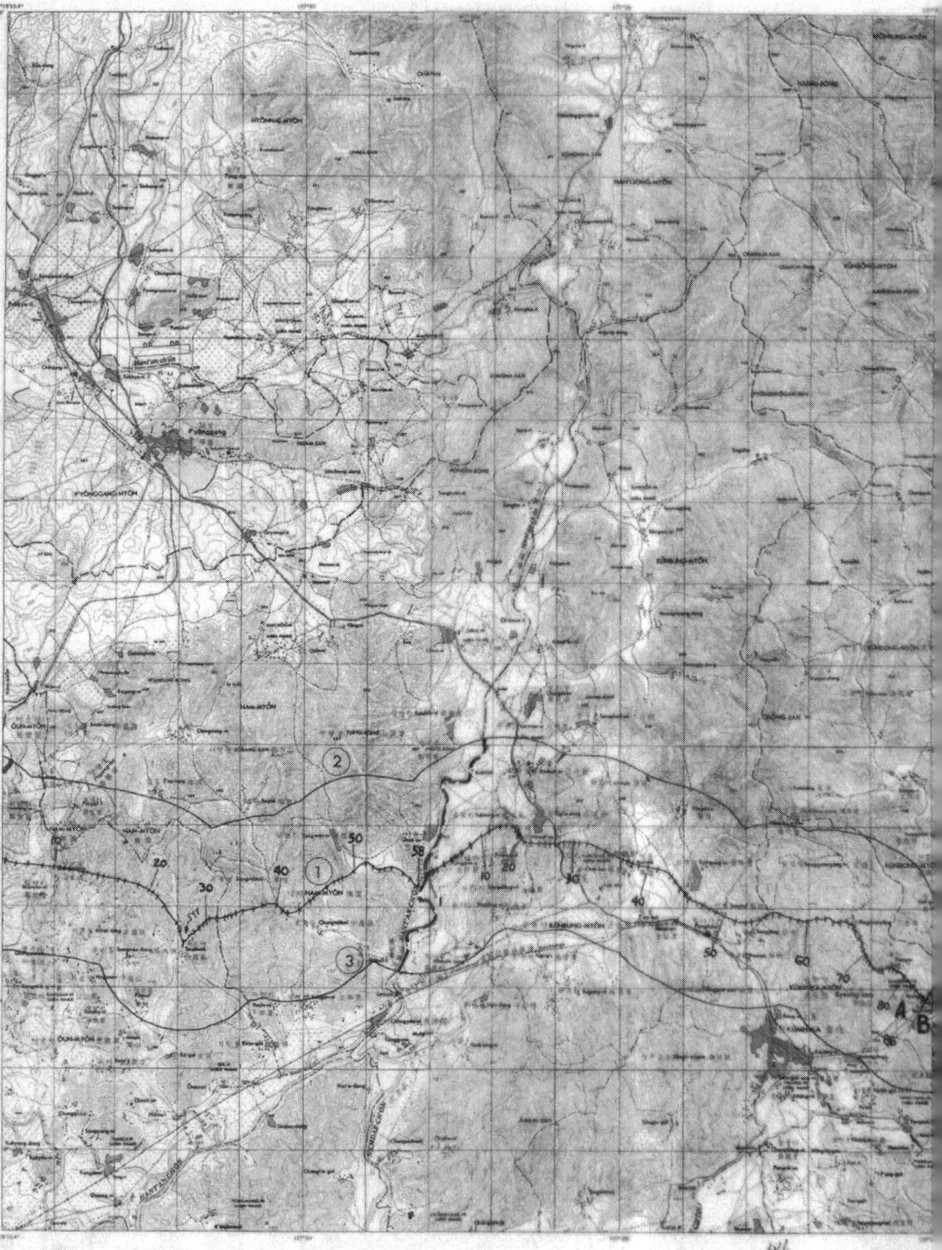

LEGEND.......SEE FOLDER INSIDE FRONT COVER

① 군사 분계선 The Military Demarcation Line 軍事分界線

② 비무장 지대의 북 경계선 The Northern Boundary of the Demilitarized Zone 非軍事區的北線

③ 비무장 지대의 남 경계선 The Southern Boundary of the Demilitarized Zone 非軍事區的南線

AUTHENTICATED

1 APRIL 54
DATE

평강
P'YŎNGGANG
平康

MAJ. HEWE
STAFF OFFICER

COL. RI
STAFF OFF.

첨부한 지도 제1도의6
군사 분계선 및
비무장 지대의 북 경계선과 남 경계선
(제2항 및 제3항을 보라)

MAP 1-6
THE MILITARY DEMARCATION LINE AND THE NORTHERN AND SOUTHERN BOUNDARIES OF THE DEMILITARIZED ZONE
(See Paragraphs 2 and 3)

附圖一之六
軍事分界線及非軍事區的北線與南線
(見第二款與第三款)

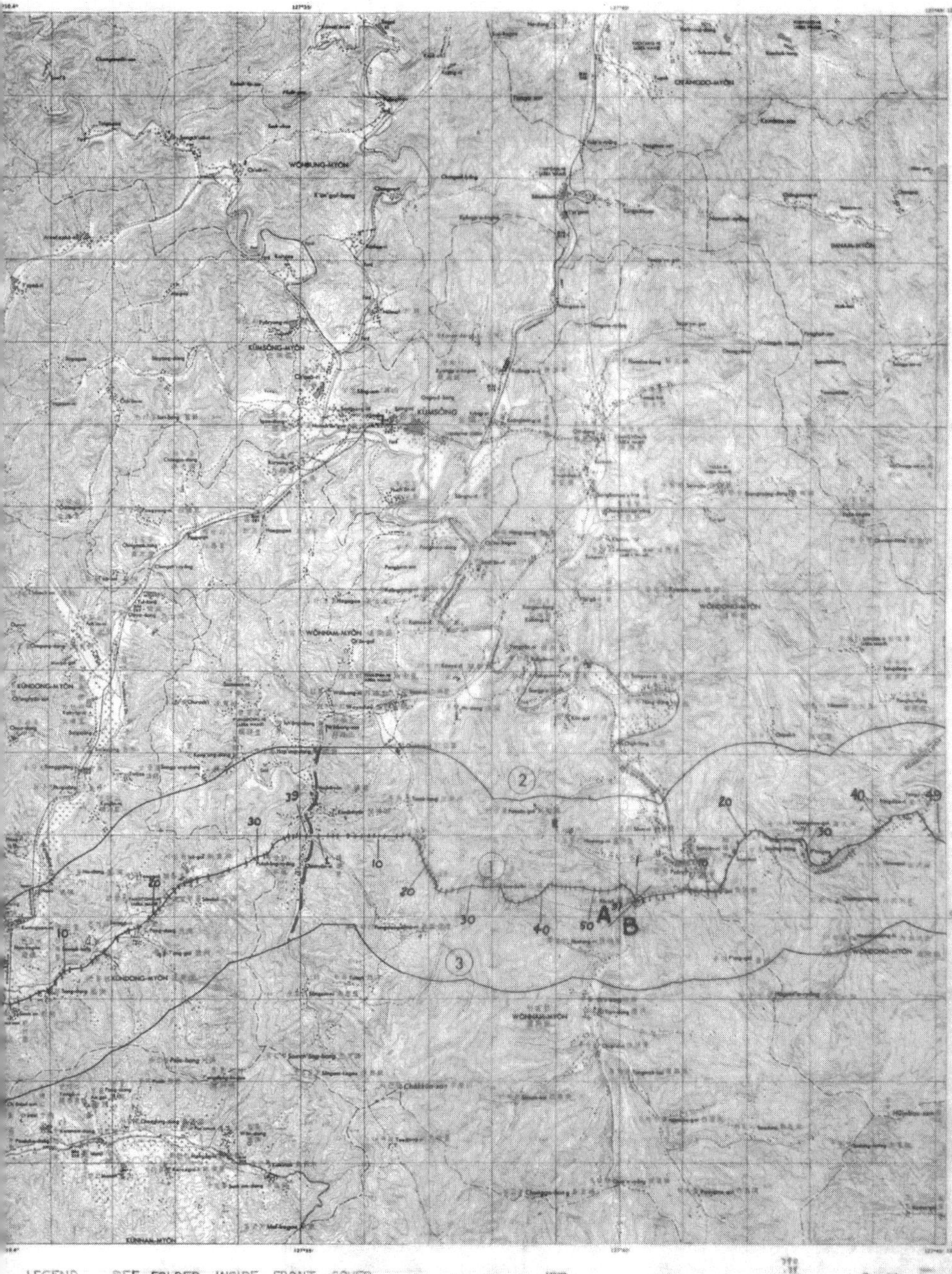

LEGEND...... SEE FOLDER INSIDE FRONT COVER

1) 군사 분계선 The Military Demarcation Line 軍事分界線

2) 비무장 지대의 북 경계선 The Northern Boundary of the Demilitarized Zone 非軍事區的北線

3) 비무장 지대의 남 경계선 The Southern Boundary of the Demilitarized Zone 非軍事區的南線

AUTHENTICATED

1 APRIL 54
DATE

금성
KŬMSŎNG
金城

MAJ HEMEYER
STAFF OFFICER
COL RI YOUNG
STAFF OFFICER

첨부한 지도 제1도의 7
군사 분계선 및
비무장 지대의 북 경계선과 남 경계선
(제2항 및 제3항을 보라)

MAP 1-7
THE MILITARY DEMARCATION LINE AND THE NORTHERN AND SOUTHERN BOUNDARIES OF THE DEMILITARIZED ZONE
(See Paragraphs 2 and 3)

附圖一之七
軍事分界線及非軍事區的北緣與
（見第二款與第三款）

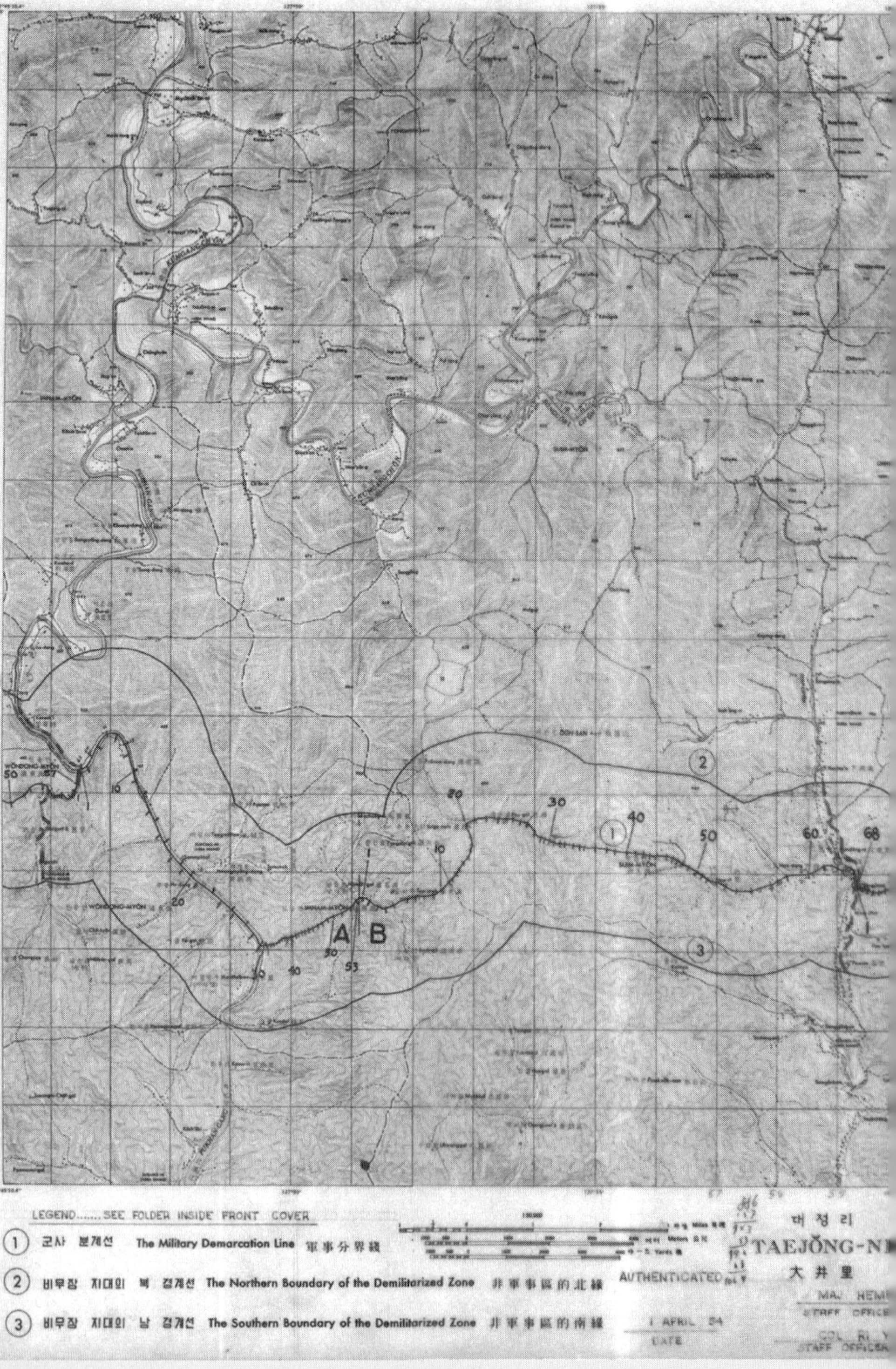

LEGEND......SEE FOLDER INSIDE FRONT COVER

① 군사 분계선 The Military Demarcation Line 軍事分界線

② 비무장 지대의 북 경계선 The Northern Boundary of the Demilitarized Zone 非軍事區的北線

③ 비무장 지대의 남 경계선 The Southern Boundary of the Demilitarized Zone 非軍事區的南線

대정리
TAEJŎNG-NI
大井里

AUTHENTICATED

1 APRIL 54
DATE

MAJ. HEM
STAFF OFFICE

COL. RI Y
STAFF OFFICER

첨부한 지도 제1도의8
군사 분계선 및
무장 지대의 북 경계선과 남 경계선
(제2항 및 제3항을 보라)

MAP 1-8
THE MILITARY DEMARCATION LINE AND THE NORTHERN AND SOUTHERN BOUNDARIES OF THE DEMILITARIZED ZONE
(See Paragraphs 2 and 3)

附圖一之八
軍事分界線及非軍事區的北緣與南緣
(見第二款與第三款)

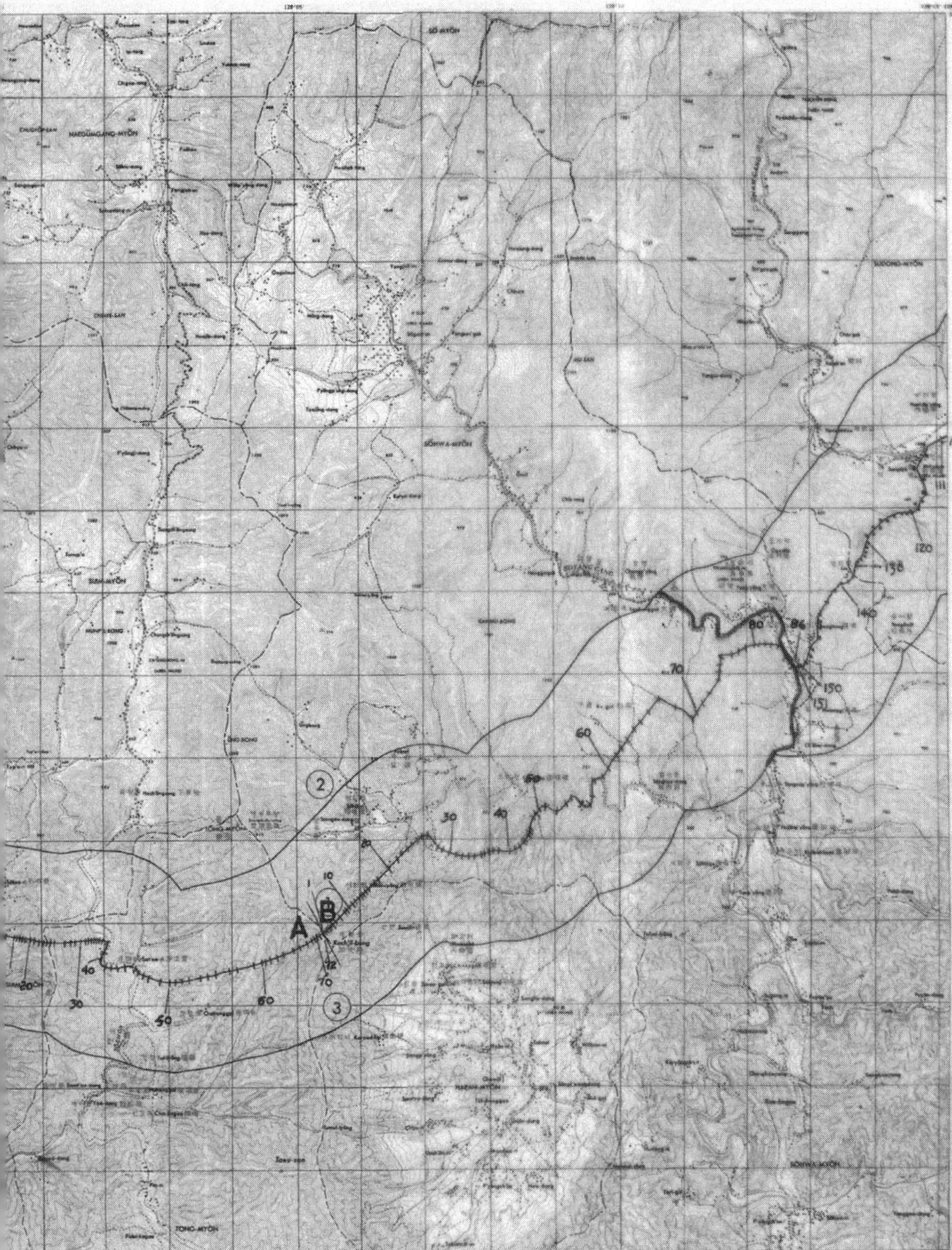

LEGEND........SEE FOLDER INSIDE FRONT COVER

1) 군사 분계선 The Military Demarcation Line 軍事分界線

2) 비무장 지대의 북 경계선 The Northern Boundary of the Demilitarized Zone 非軍事區的北緣

3) 비무장 지대의 남 경계선 The Southern Boundary of the Demilitarized Zone 非軍事區的南緣

만대리
MANDAE-RI
萬垈里

AUTHENTICATED

MAJ HEMEYEYER
STAFF OFFICER UNC

1 APRIL 54
DATE

COL RI YOUNG
STAFF OFFICER KPA/CPV

첨부한 지도 제1도의9
군사 분계선 및
비무장 지대의 북 경계선과 남 경계선
(제2항 및 제3항을 보라)

MAP 1-9
THE MILITARY DEMARCATION LINE AND THE NORTHERN AND SOUTHERN BOUNDARIES OF THE DEMILITARIZED ZONE
(See Paragraphs 2 and 3)

附圖一之九
軍事分界線及非軍事區的北緣與
(見第二款與第三款)

① 군사 분계선 The Military Demarcation Line 軍事分界線

② 비무장 지대의 북 경계선 The Northern Boundary of the Demilitarized Zone 非軍事區的

③ 비무장 지대의 남 경계선 The Southern Boundary of the Demilitarized Zone 非軍事區的

LEGEND
SEE FOLDER INSIDE FRONT COVER

AUTHENTICATED
1 APRIL 54
DATE

MAJ HEMEYER UNC
STAFF OFFICER UNC

COL RI YOUNG
STAFF OFFICER KPA/CPV

고 성
KOSŎN
高城

첨부한 지도 제2도
군사정전위원회
감독하의 한강구
구역
(제5항을 보라)

MAP 2
THE AREA OF THE HAN RIVER ESTUARY UNDER THE SUPERVISION OF THE MILITARY ARMISTICE COMMISSION
(See Paragraph 5)

附圖二
軍事停戰委員會監督下的漢江口區域
(見第五款)

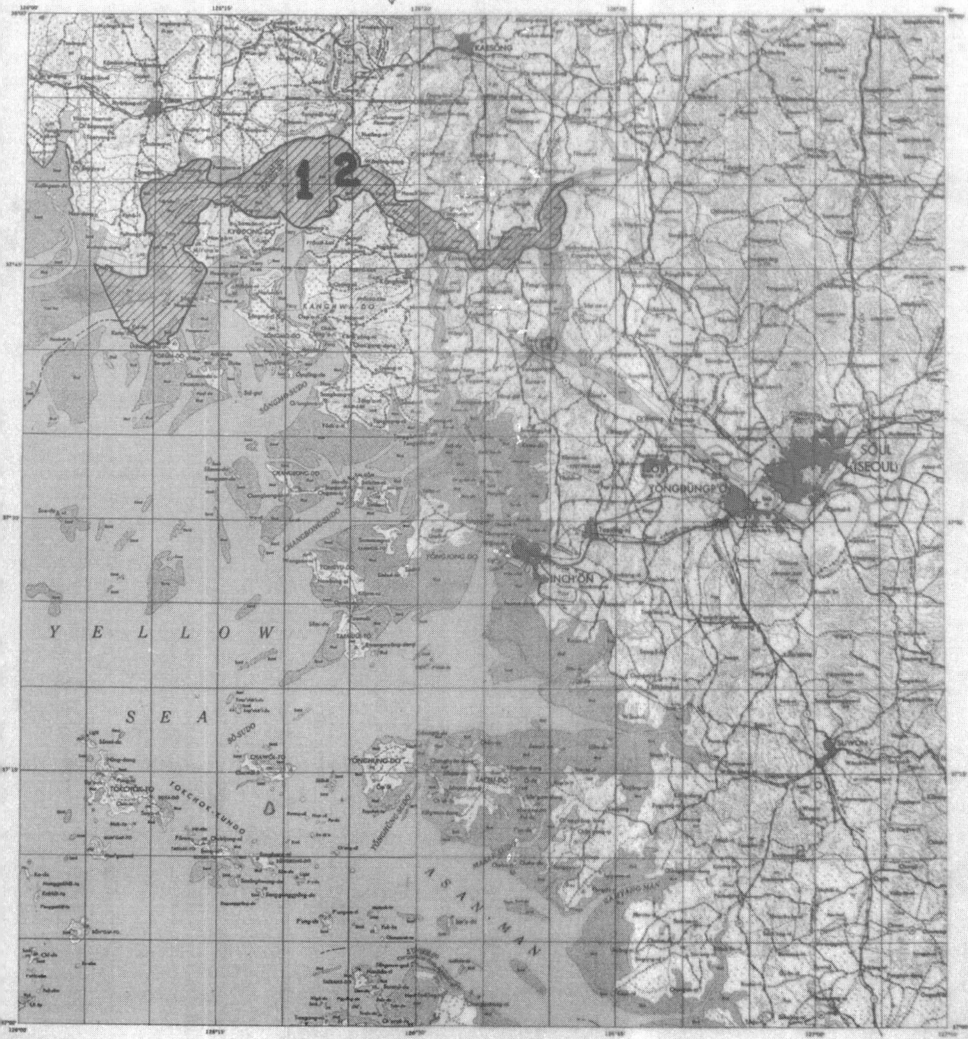

서울
SŎUL(SEOUL)
京城

LEGEND
SEE FOLDER INSIDE FRONT COVER

AUTHENTICATED

MAJ HEMEYER USA
STAFF OFFICER UNC

1 APRIL 54
DATE

COL RI YOUNG
STAFF OFFICER KPA/CPV

첨부한 지도 제3호

한국 서부연해섬들의 통제

(제13항 ㄴ목을 보라)

황해도와 경기도의 도계선 (가—나선)북쪽과 서쪽에있는 모든섬중에서 하기한 다섯(5)도서군들을 제외한 기타 모든 섬들은 조선인민군 최고사령관과 중국인민지원군 사령원의 군사통제하에 둔다.

1. 백 령 도 북위 37도58분 동경 124도40분
2. 대 청 도 북위 37도50분 동경 124도42분
3. 소 청 도 북위 37도46분 동경 124도46분
4. 연 평 도 북위 37도38분 동경 125도40분
5. 우 도 북위 37도36분 동경 125도58분

상기다섯(5)도서군들은 국제련합군 총사령관의 군사통제하에 남겨둔다. 한국 서부해안에 있어서 상기 경계선 이남에있는 모든 섬들은 국제련합군 총사령관의 군사통제하에 남겨둔다.

(주1) 상기계선 (가—나선)의 목적은 다만 한국 서부연해섬들의 통제를 표시하는것이다. 이선은 아무런 다른 의의가 없으며 또한 이에 다른 의의를 첨부하지도 못한다.

(주2) 각도서군들을 둘러싼 장방형의 구획의 목적은 다만 국제련합군 총사령관의 군사통제하에 남겨두는 각도서군들을 표시하는 것이다. 이러한 장방형의 구획은 아무런 다른 의의가 없으며 또한 이에 다른 의의를 첨부하지도 못한다.

MAP 3

CONTROL OF COASTAL ISLANDS ON THE WEST COAST OF KOREA

(See Sub-paragraph 13b)

All the islands lying to the north and west of the provincial boundary lin (line A-B) between HWANGHAE-DO and KYONGGI-DO shall be under th military control of the Supreme Commander of the Korean People's Army an the Commander of the Chinese People's Volunteers, except the following fiv island groups:

1. PAENGYONG-DO Lat 37°58'N Long 124°40'E
2. TAECHONG-DO Lat 37°50'N Long 124°42'E
3. SOCHONG-DO Lat 37°46'N Long 124°46'E
4. YONPYONG-DO Lat 37°38'N Long 125°40'E
5. U-DO Lat 37°36'N Long 125°58'E

The above five island groups shall remain under the military control of th Commander-in-Chief, United Nations Command. All the islands on the we coast of Korea lying south of the above-mentioned boundary line shall re main under the military control of the Commander-in-Chief, United Nation Command.

Notes: (1) The purpose of the line A-B is solely to indicate the control c coastal islands on the west coast of Korea. This line has no other signif cance and none shall be attached thereto.

(2) The rectangles which inclose the island groups are for the sol purpose of indicating island groups which shall remain under the militar control of the Commander-in-Chief, United Nations Command. Thes rectangles have no other significance and none shall be attached theret

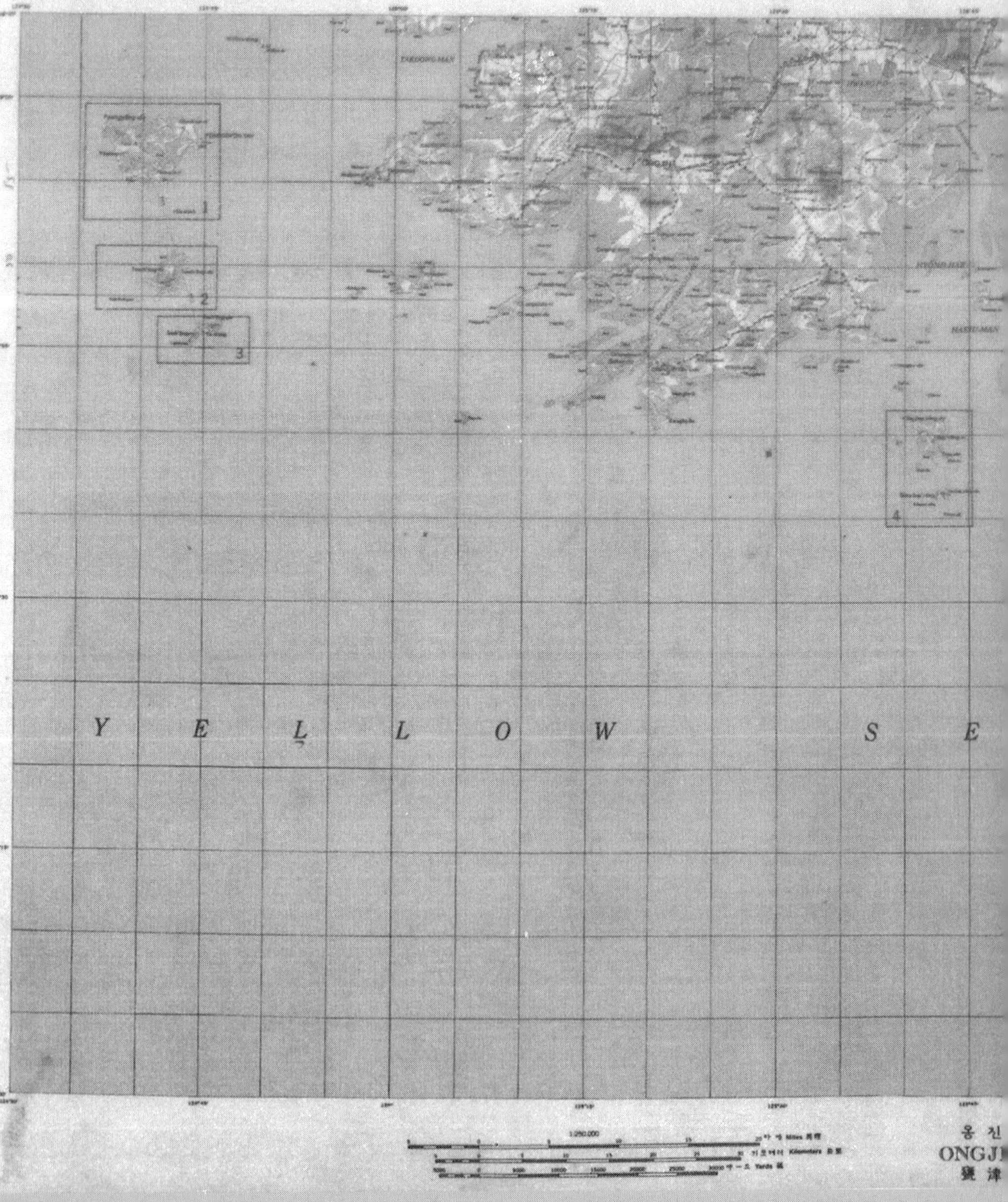

附圖三

韓國西岸沿海島嶼的控制

（見第十三款丑項）

於黃海道與京畿道道界（甲 — 乙線）以北及以
切島嶼，除下列五島群外，均置於朝鮮人民軍最
官與中國人民志願軍司令員的軍事控制之下：

翎島	北緯三七度五八分	東經一二四度四〇分
青島	北緯三七度五〇分	東經一二四度四二分
青島	北緯三七度四六分	東經一二四度四六分
坪島	北緯三七度三八分	東經一二五度四〇分
島	北緯三七度三六分	東經一二五度五八分

此五島群留置聯合國軍總司令的軍事控制之下，
岸位於上述界線以南的一切島嶼均留置聯合國軍
的軍事控制之下。

（註一）上述界線（甲 — 乙線）的目的僅為標明韓國西岸沿海島嶼的控制，此線並無其他意義，亦不得附加以其他意義。

（註二）圍繞各島群的長形方框的目的僅為標明留置聯合國軍總司令的軍事控制下的各島群，此等長形方框並無其他意義，亦不得附加以其他意義。

1:250,000

첨부된 지도 제4호
주요 교통선
(제13항 ㄱ목을 보라)
중립국시찰소조가 중립국감독위원회 본부와 제43항에 열거한 출입항구들 사이에 충분한 통행상 편리를 가져야할 통로

MAP 4
MAIN LINES OF COMMUNICATION
(See Sub-paragraph 13g)

Routes over which Neutral Nations Inspection Teams shall be afforded full convenience of movement between the headquarters of the Neutral Nations Supervisory Commission and the ports of entry enumerated in Paragraph 43.

附圖四
主要交通線
（見第十三款午項）

中立國視察小組往返於中立國監察委員會總部與第四十三款所開列的口岸時應有充分通行便利的通道。

한국
KOREA
韓國

1:1,000,000

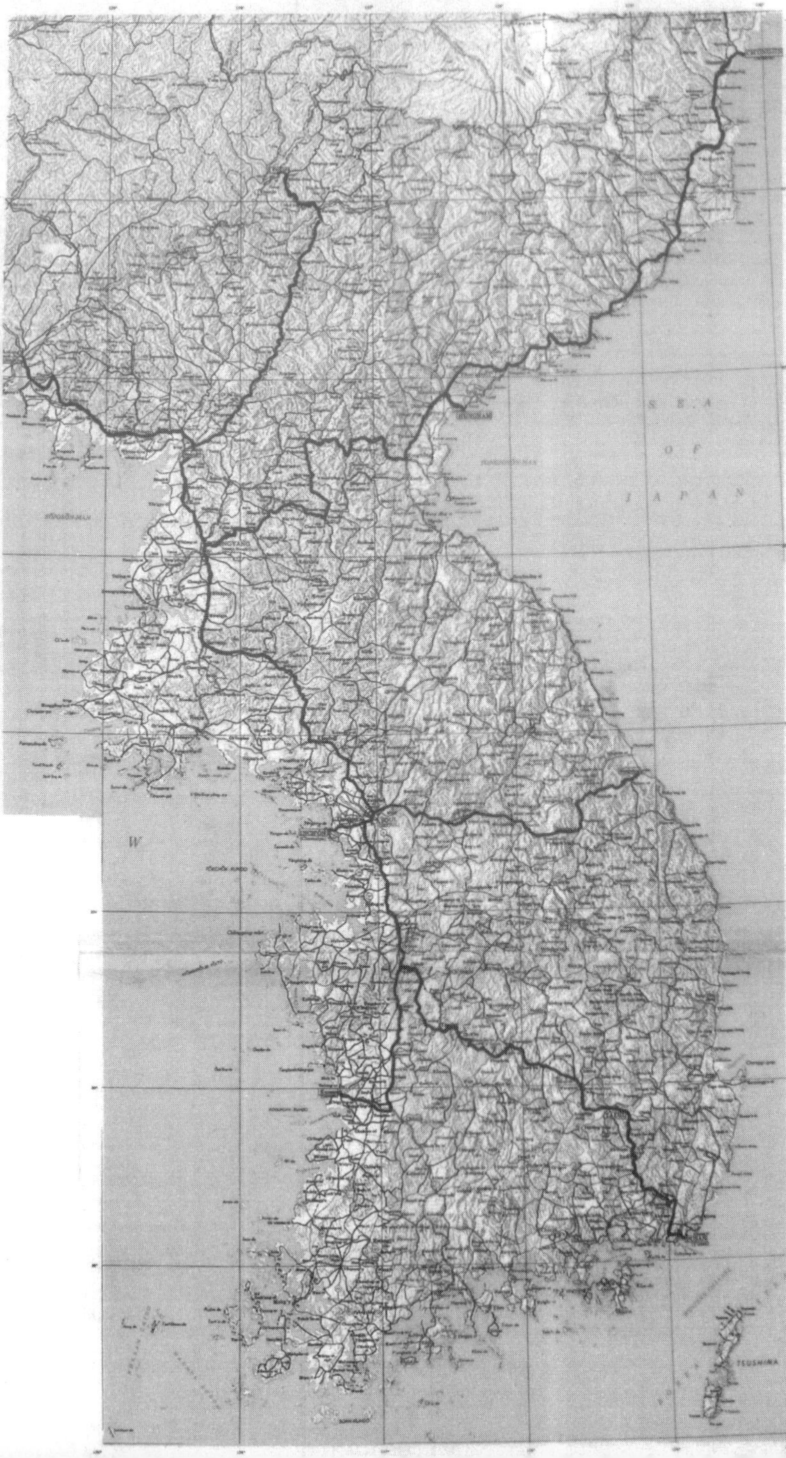

첨부한 지도 제5도의1

신의주 출입항

(제43항을 보라)

교대인원, 작전비행기, 장갑차량, 무기 및 탄약이 한국으로 들어오는 구역과 중립국시찰소조가 충분한 통행상 편리를 가지는 구역을 표시한다

MAP 5-1

PORT OF ENTRY OF SINUIJU

(See Paragraph 43)

Showing the area through which rotation personnel, combat aircraft, armored vehicles, weapons, and ammunition may be introduced into Korea and within which the Neutral Nations Inspection Team shall have full convenience of movement.

附圖五之一

新義州口岸

（見第四十三款）

表示輪換人員、作戰飛機、裝甲車輛、武器與彈藥得進入韓國的區域與中立國視察小組有充分通行便利的區域。

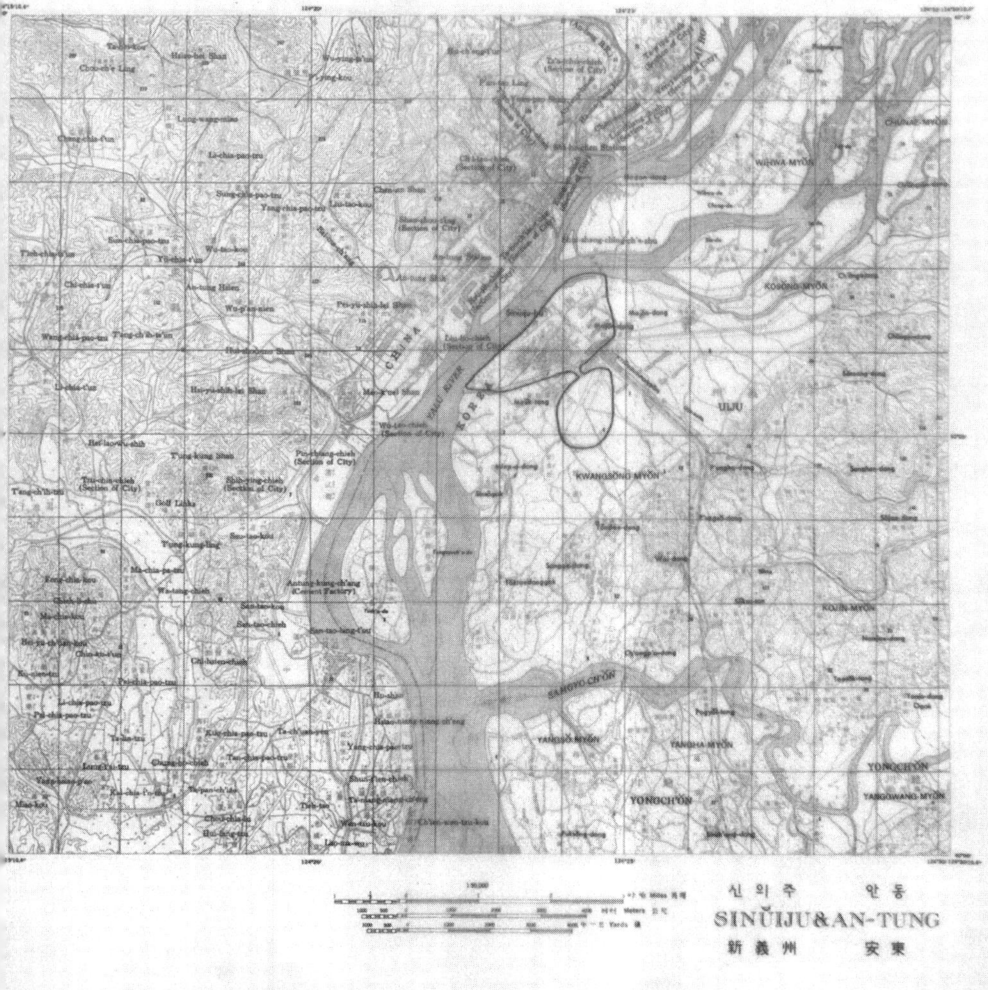

신의주 안동

SINŬIJU & AN-TUNG

新義州 安東

첨부한 지도 제5도의2

청 진 출입항

(제43항을 보라)

윤환인원, 작전비행기, 장갑차량, 무기 및 탄약이 한국으로 들어오는 구역과 중립국시찰소조가 충분한 통행상 편리를 가지는 구역을 표시한다

MAP 5-2

PORT OF ENTRY OF CHONGJIN

(See Paragraph 43)

Showing the area through which rotation personnel, combat aircraft, armored vehicles, weapons, and ammunition may be introduced into Korea and within which the Neutral Nations Inspection Team shall have full convenience of movement.

附圖五之二

清津口岸

（見第四十三款）

表示輪換人員、作戰飛機、裝甲車輛、武器與彈得進入韓國的區域與中立國視察小組有充分通行便利區域。

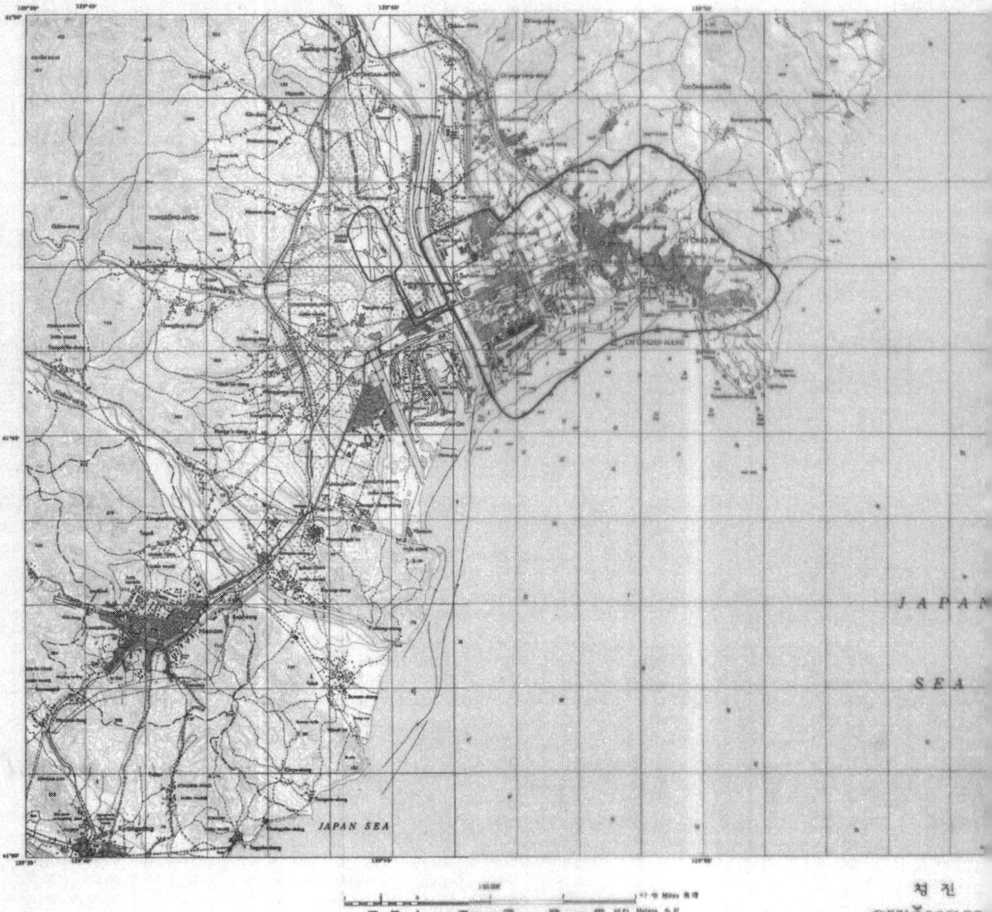

청 진

CH'ŎNGJI

清津

첨부한 지도 제5도의3

흥남 출입항

(제43항을 보라)

교환인원, 작전비행기, 장갑차량, 무기 및 탄약이 한국으로 드러오는 구역과 중립국시찰소조가 충분한 통행상 편리를 가지는 구역을 표시한다

MAP 5-3
PORT OF ENTRY OF HUNGNAM
(See Paragraph 43)

Showing the area through which rotation personnel, combat aircraft, armored vehicles, weapons, and ammunition may be introduced into Korea and within which the Neutral Nations Inspection Team shall have full convenience of movement.

附圖五之三

興南口岸

（見第四十三款）

表示輪換人員、作戰飛機、裝甲車輛、武器與彈藥得進入韓國的區域與中立國視察小組有充分通行便利的區域．

함흥
HAMHŬNG
咸興

첨부한 지도 제5도의4

만 포 출입항

(제43항을 보라)

윤환인원, 작전비행기, 장갑차량, 무기 및 탄약이 한국으로 드러오는 구역과 중립국시찰소조가 충분한 통행상 편리를 가지는 구역을 표시한다

MAP 5-4

PORT OF ENTRY OF MANPO

(See Paragraph 43)

Showing the area through which rotation personnel, combat aircraft, armored vehicles, weapons, and ammunition may be introduced into Korea and within which the Neutral Nations Inspection Team shall have full convenience of movement.

附圖五之四

滿浦口岸

（見第四十三款）

表示輪換人員、作戰飛機、裝甲車輛、武器與彈得進入韓國的區域與中立國視察小組有充分通行便利區域。

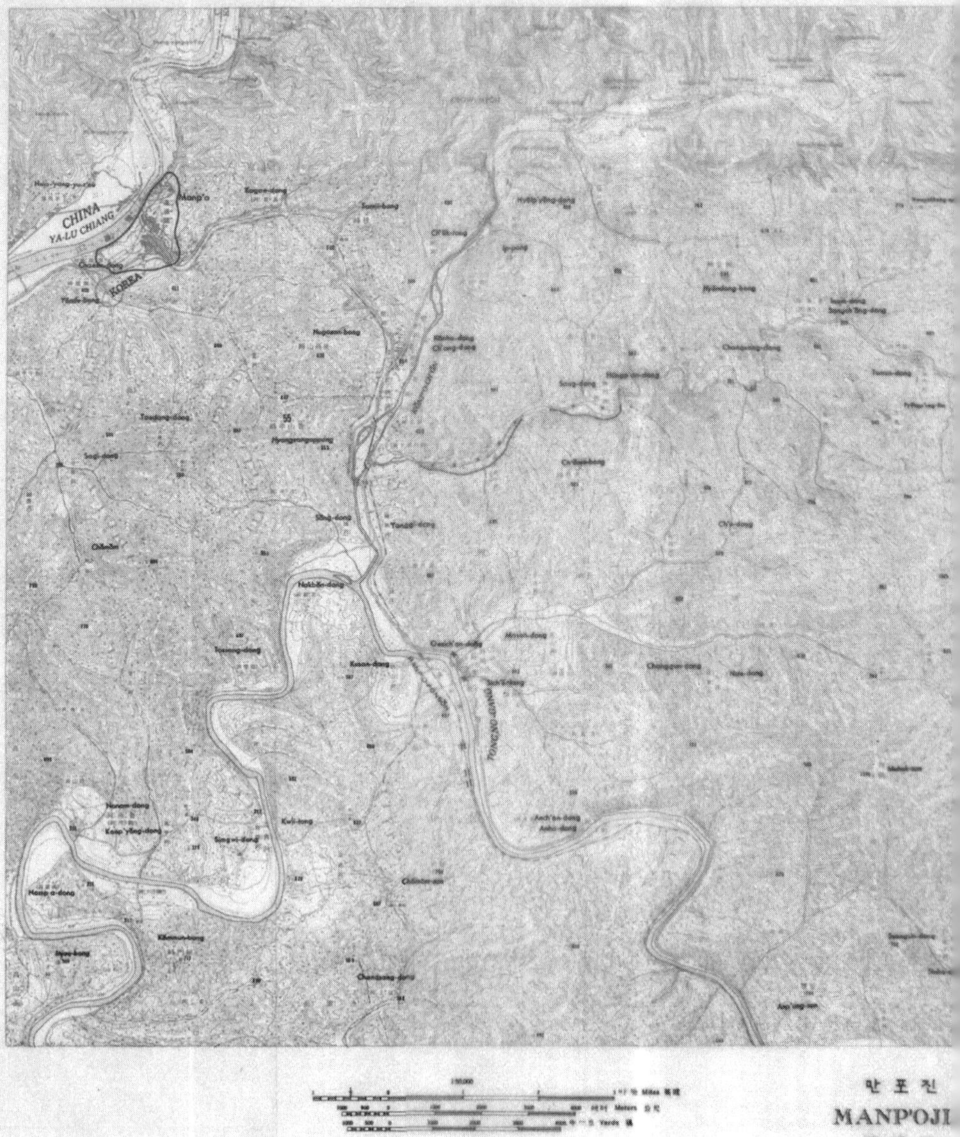

만 포 진

MANP'OJI

滿浦津

첨부한 지도 제5도의5

신안주 출입항

(제43항을 보라)

교환인원, 작전비행기, 장갑차량, 무기 및 탄약이 한국으로 들어오는 구역과 중립국시찰소조가 충분한 통행상 편리를 가지는 구역을 표시한다

MAP 5-5
PORT OF ENTRY OF SINANJU

(See Paragraph 43)

Showing the area through which rotation personnel, combat aircraft, armored vehicles, weapons, and ammunition may be introduced into Korea and within which the Neutral Nations Inspection Team shall have full convenience of movement.

附圖五之五

新安州口岸

（見第四十三款）

表示輪換人員、作戰飛機、裝甲車輛、武器與彈藥得運入韓國的區域與中立國視察小組有充分通行便利的區域。

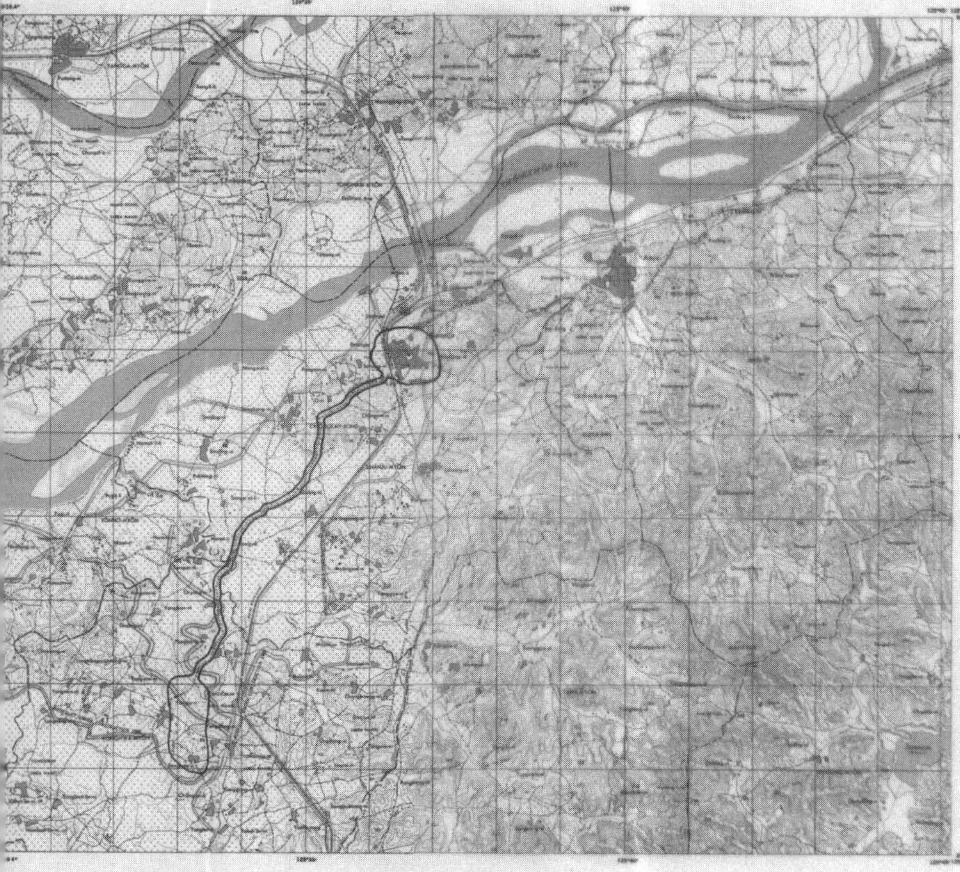

안주

ANJU

安州

첨부한 지도 제5도의6

인 천 출입항

(제43항을 보라)

윤환인원, 작전비행기, 장갑차량,
무기 및 탄약이 한국으로 들어
오는 구역과 중립국시찰소조가
충분한 통행상 편리를 가지는
구역을 표시한다

MAP 5-6

PORT OF ENTRY OF INCHON

(See Paragraph 43)

Showing the area through which rotation per
aircraft, armored vehicles, weapons, and ammu
introduced into Korea and within which the N
Inspection Team shall have full convenience of m

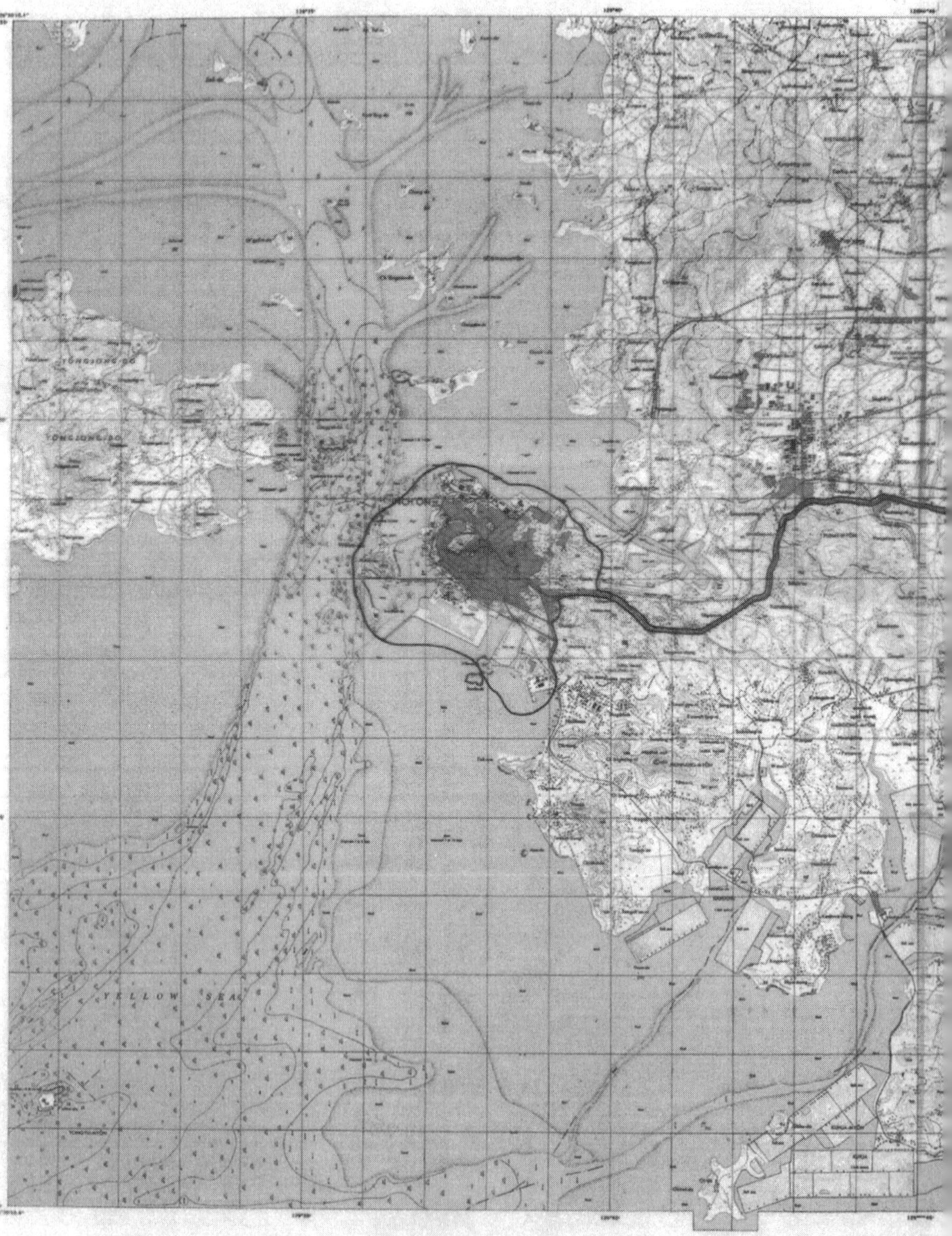

附圖五之六

仁川口岸

（見第四十三款）

標示輪換人員、作戰飛機、裝甲車輛、武器與彈藥得進入韓國的區域與中立國監察小組有充分通行便利的區域．

인 천
INCH'ŎN
仁 川

첨부한 지도 제5도의7

대 구 출입항

(제4 3항을 보라)

교환인원, 작전비행기, 장갑차량, 무기 및 탄약이 한국으로 들어오는 구역과 중립국시찰소조가 충분한 통행상 편리를 가지는 구역을 표시한다

MAP 5-7

PORT OF ENTRY OF TAEGU

(See Paragraph 43)

Showing the area through which rotation personnel, combat aircraft, armored vehicles, weapons, and ammunition may be introduced into Korea and within which the Neutral Nations Inspection Team shall have full convenience of movement.

附 圖 五 之 七

大邱口岸

（見第四十三款）

表示輪換人員、作戰飛機、裝甲車輛、武器與彈
得運入韓國的區域與中立國視察小組有充分通行便利
區域.

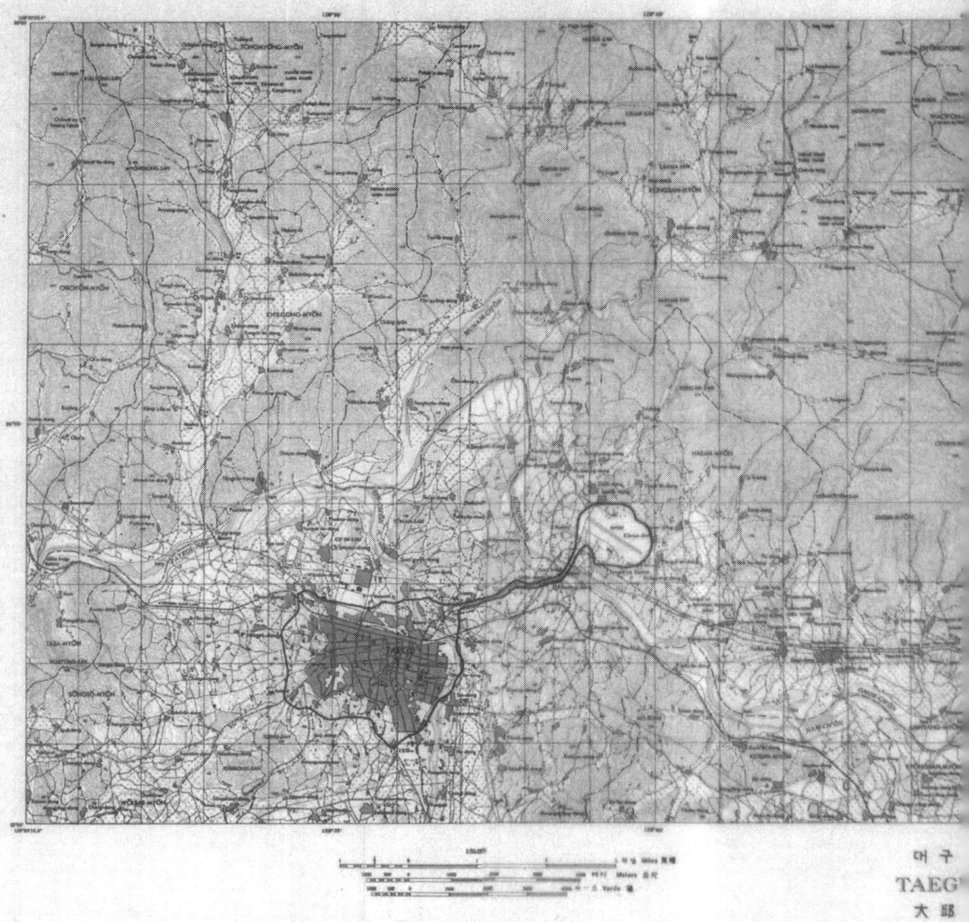

대 구
TAEG
大 邱

첨부한 지도 제5도의8

부 산 출입항

(제43항을 보라)

윤환인원, 작전비행기, 장갑차량, 무기 및 탄약이 한국으로 드려오는 구역과 중립국시찰소조가 충분한 통행상 편리를 가지는 구역을 표시한다.

MAP 5-8

PORT OF ENTRY OF PUSAN

(See Paragraph 43)

Showing the area through which rotation personnel, combat aircraft, armored vehicles, weapons, and ammunition may be introduced into Korea and within which the Neutral Nations Inspection Team shall have full convenience of movement.

附圖五之八

釜山口岸

（見第四十三款）

表示輪換人員、作戰飛機、裝甲車輛、武器與彈藥得進入韓國的區域與中立國視察小組有充分通行便利的區域。

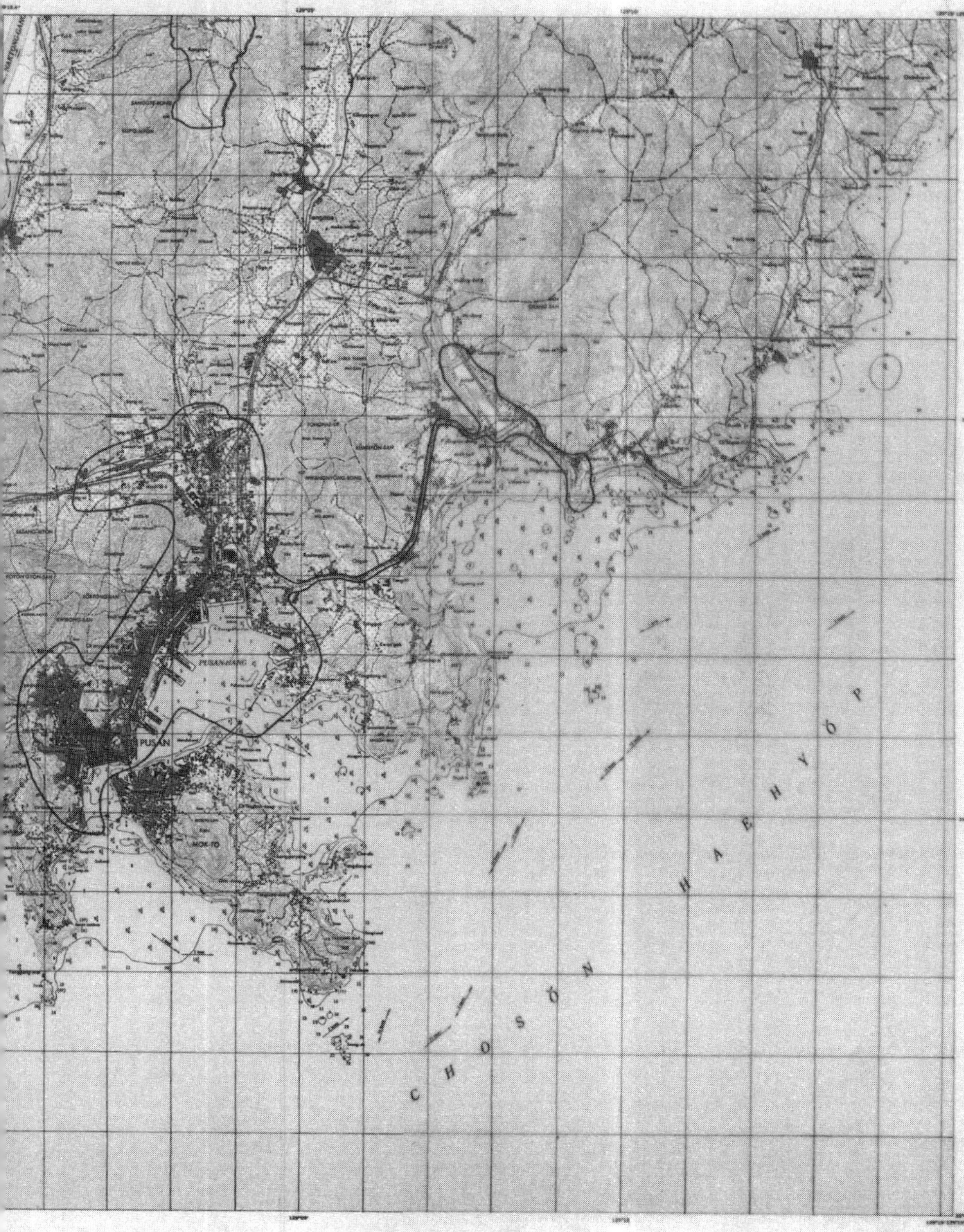

부산

PUSAN

釜山

첨부한 지도 제5호의9

강 릉 출입항

(제43항을 보라)

륜환인원, 작전비행기, 장갑차량, 무기 및 탄약이 한국으로 드러 오는 구역과 중립국시찰소조가 충분한 통행상 편리를 가지는 구역을 표시한다

MAP 5-9
PORT OF ENTRY OF KANGNUNG
(See Paragraph 43)

Showing the area through which rotation personnel, combat aircraft, armored vehicles, weapons, and ammunition may be introduced into Korea and within which the Neutral Nations Inspection Team shall have full convenience of movement.

附圖五之九

江陵口岸

（見第四十三款）

表示輪換人員、作戰飛機、裝甲車輛、武器與彈藥得進入韓國的區域與中立國視察小組有充分通行便利的區域。

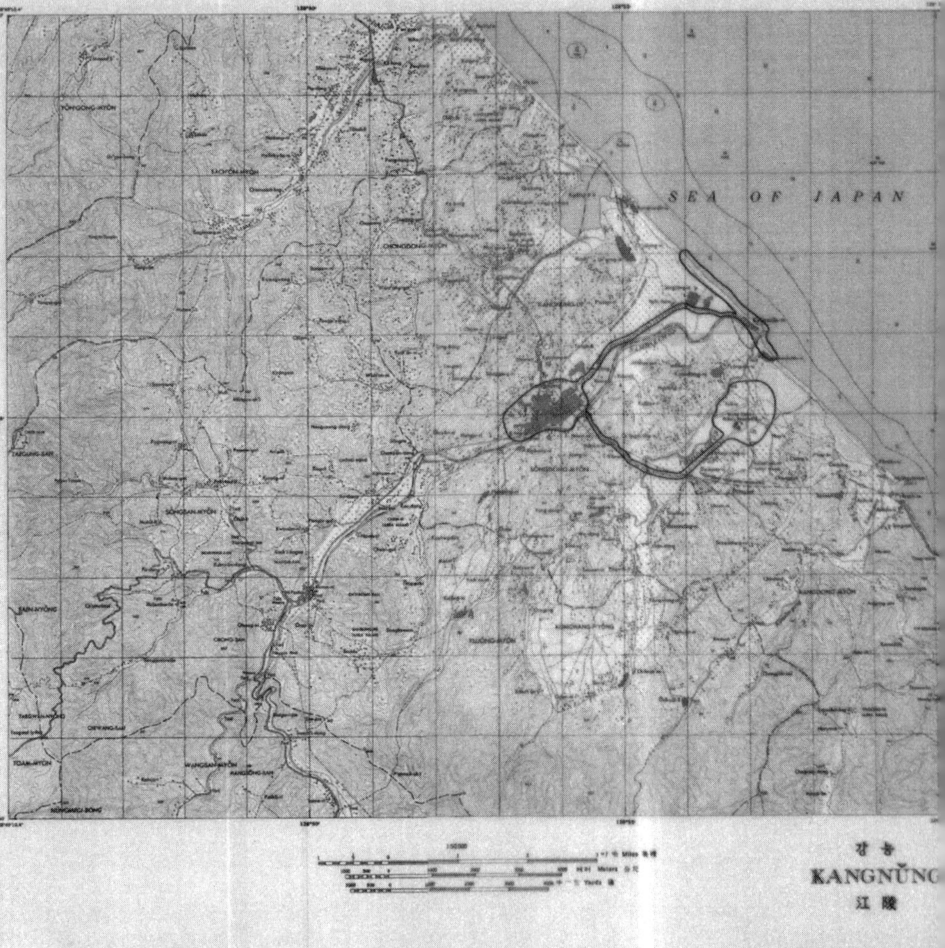

강 릉
KANGNŬNG
江陵

…한 지도 제5도의10
군 산 출입항
(제43항을 보라)
…원, 작전비행기, 장갑차량,
및 탄약이 한국으로 도입
구역과 중립국시찰소조가
통행상 편리를 가지는
표시한다

MAP 5-10
PORT OF ENTRY OF KUNSAN
(See Paragraph 43)

Showing the area through which rotation personnel, combat aircraft, armored vehicles, weapons, and ammunition may be introduced into Korea and within which the Neutral Nations Inspection Team shall have full convenience of movement.

附圖五之十

群山口岸

（見第四十三款）

表示輪換人員、作戰飛機、裝甲車輛、武器與彈藥得進入韓國的區域與中立國視察小組有充分通行便利的區域。

군산
KUNSAN
群山

4. INDEX TO AGREEMENTS SUBSEQUENT TO THE SIGNING OF THE ARMISTICE AGREEMENT
(Revised 15 April 2004)

1. Rules for Civil Shipping in the Han River Estuary and Related Matters
2. Definition of the term "Combat Aircraft"
3. Understanding on Administrative Details for the Delivery and Reception of Bodies of Military Personnel of Both Sides
4. Agreement on the MAC Headquarters Area, its Security and its Construction
5. Use of Public Address System in the Headquarters Area
6. Understanding on Administrative Details for Assisting the return of displaced Civilians and the Proceeding of Civilians of Foreign Nationality to Territory Under the Control of the other side
7. Amendment to Agreement on Entry into and Departure from the Demilitarized Zone by Civilians
8. Agreement on Credentials, Distinctive Insignia and Markings for the MAC, its Agencies and the Personnel of such Organizations
9. Revised General Rules for the Control, Organization, Operation and Support of the Joint Observer Teams
10. Chronology of Negotiation for Reduction of Joint Observer Teams and Revised General Rules for Joint Observer Teams
11. Understanding concerning the Repair and Maintenance of the MDL Markers on both banks of the Imjin River, North Han River and Kumsong River Where the MDL runs down the center of the Rivers
12. Designation of "A" and "B" Sectors of MDL and Numbering

4. 정전협정 체결 이후 추가 합의사항 목록(2004년 4월 15일 수정)

*핵심사항은 선별하여 게재하였습니다.
1, 9, 10, 17, 21, 28, 29항

1. 한강하구에서의 민용선박 항행에 대한 규칙 및 관계사항
2. "작전 비행기"의 정의
3. 쌍방 군사 인원 시체 인도 인수에 관한 행정상 세목의 양해
4. 군사정전위원회 본부구역, 본부구역의 안전 및 본부구역의 수축에 관한 합의
5. 본부구역내에서의 확성기 사용
6. 실향 사민의 귀향과 외국적의 사민이 상대방 통제 지역으로 가는 것을 협조함에 관한 행정상 세목의 양해

7. 사민의 비무장지대 출입에 관한 합의의 수정

8. 군사정전위원회 및 그 종속기관과 각 해당기관 인원의 증명서 휘장 및 식별 표식에 관한 합의
9. 공동감시소조의 관리, 조직, 사업 및 보급에 관한 수정 총칙
10. 공동감시소조 수의 축소 및 수정 개칙을 위한 협의 연대표
11. 군사분계선이 임진강, 북한강 및 금성강등의 중앙을 통과하는 강 양 연안에 있는 군사분계선 표식물의 보수 및 유지에 관한 양해

12. 군사분계선 "갑"구와 "을"구의 지정과 표식물들의

System and Marker Designs
13. Agreement on Chinese Linguists on JOT Meetings

14. Agreement on Intermission of MAC & Sec Meetings Every three hours
15. Agreement concerning the Logistical Support and Expense of the MAC, the NNSC, and NNRC, their Respective subordinate bodies and other personnel
16. Provisions regarding the attendance of news media Representatives in the Demilitarized Zone
17. Procedures for supervising and reporting the arrivals in and departures from Korea of Military Personnel and Replacement Items
18. Operating Procedure for Joint Duty Office
19. Tentative Rules of Procedure of the Military Armistice Commission
20. Addenda to Tentative Rules of Procedure of the Military Armistice Commission
- Agreement on calling Military Armistice Commission Meetings
21. Agreement on reduction of NNITS
22. Agreements on use of Chinese Linguists at Secretary Meetings

23. Agreement on use of Military Police as Civil Police in the DMZ and use of types of weapons

24. Agreement on number of copies of communications to be passed between Sr Mbrs of both sides
25. Agreement that the MDL has been marked
26. Agreement on calling Secretaries' Meetings
27. Procedures for UNC-KPA General Officer-Level Dialogue on Armistice Issues

번호 배당 방법 및 표식물의 규격

13. 중국어 언어인원의 공동감시소조 회의 참가 여부에 관한 합의
14. 군사정전위원회 및 비서장회의시 세 시간마다 휴회하는 데 대한 합의
15. 군사정전위원회, 중립국감독위원회, 중립국 송환위원회 및 그의 각 종속 기관과 기타 인원에게 대한 보급과 비용에 관한 합의
16. 비무장지대내에서의 보도 기관 대표 참석에 관한 규정
17. 군사인원의 한국에의 도착과 한국으로부터의 이거와 물건의 교체를 감독하며 보고하는 데 관한 절차
18. 공동일직장교 사무실의 운영 절차
19. 군사정전위원회의 잠정적 절차 규정
20. 군사정전위원회의 잠정적 절차 규정에 대한 증보
    - 군사정전위원회 회의 소집에 대한 합의

21. 중립국 시찰소조 축소에 관한 합의
22. 비서장회의에 있어서의 중국어 언어 인원의 사용에 관한 합의
23. 비무장지대내에서 군사 경찰을 민사 경찰로서 사용하는 데 관한 합의 및 민사 경찰이 휴대할 수 있는 무기의 종류에 관한 합의
24. 쌍방 수석위원간에 왕래되는 통신문 사본 수에 관한 합의
25. 군사분계선이 표식된데 관한 합의
26. 비서장회의 소집에 관한 합의
27. 정전문제에 관한 국제연합군과 조선인민군간의 장군급 대화를 위한 절차

28. Agreement between the United Nations Command and the Korean Peoples Army on Opening Some Area of the DMZ (West Side, 17 November 2000)
29. Agreement between the United Nations Command and the Korean Peoples Army on Opening Some Area of the DMZ (East Side, 12 September 2002)
30. Distribution List

1. RULES FOR CIVIL SHIPPING IN THE HAN RIVER ESTUARY AND RELATED MATTERS

1. In accordance with the provision of Paragraph 5, Article I of the Armistice Agreement, these rules for civil shipping in the Han River Estuary are prescribed.
2. Both sides shall acknowledge the Demilitarized State of the Han River Estuary Area as indicated on Map 2 attached to the Armistice Agreement. The contact line between water and land at high tide shall be made the boundary line between the Han River Estuary Area and the area under the military control of either side.
3. The Commanders of the opposing sides shall have suitable markers erected at estuaries and ports where the area under their respective control borders on the Han River Estuary Area.
4. All provisions of the Armistice Agreement which apply to the Demilitarized Zone shall also apply to the Han River Estuary Area, with the exception of the provisions relating to the fixing of the Military Demarcation Line and the provisions restricting civilians entering the Demilitarized Zone set forth in Paragraphs 9, 10, and

28. 비무장지대 일부구역 개방에 대한 국제연합군과 조선인민군간 합의서 (서측—2000년 11월 17일)

29. 비무장지대 일부구역 개방에 대한 국제연합군과 조선인민군간 합의서 (동측—2002년 9월 12일)

30. 배부처 일람표

1. 한강 하구에서의 민용선박 항행에 관한 규칙 및 관계 사항

1. 정전협정 제1조 5항의 규정에 의하여 한강 하구에서의 민용선박 항행에 대한 본 규칙을 제정한다.

2. 쌍방은 정전협정에 첨부한 지도 제 2도에 표시한 한강 하구 수역의 비무장화를 승인한다. 한강하구 수역과 각방 군사 통제지역과의 경계선은 만조시의 수륙 접촉선으로 한다.

3. 적대 쌍방 사령관은 각기 자기 통제하의 지역이 한강 하구 수역과 인접하는 데 있는 하구 및 항구에 적당한 표식물을 설치한다.

4. 정전협정 중 군사분계선을 확정함에 관한 규정과 제9항, 제10항 및 제13항 ㄱ목에서 사민이 비무장지대에 들어가는 것을 제한하는 각항 규정을 제외하고 비무장지대에 적용되는 모든 규정은 모두 한강 하구 수역에도 적용된다.

subparagraph 13a of the Armistice Agreement.

5. The Commander of each side shall have the responsibility for exercising civil administration and relief for personnel of his side. In order to maintain order and to enforce the provisions of these Rules, each side, shall, according to its requirements, provide not more than four (4) patrol boats for the civil police and not more than twenty-four (24) civil police in the Han River Estuary Area. Arms carried by the civil police shall be restricted to pistols and rifles. Civil police shall apprehend all violators from their respective sides, and assist disabled vessels of their respective sides in reaching their respective shores.
6. Unwritten rules and customs of navigation long familiar to the people for the shipping in the Han River Estuary Area shall be respected by vessels of both sides, with the exception of those in conflict with the provisions of the Armistice Agreement and the Rules contained herein.
7. All military ships, civil ships carrying military personnel, weapons and ammunition, and Neutral Nations ships shall not enter the Han River Estuary Area except with the specific authority of the Military Armistice Commission.
8. Buoys, floats, lights, panels, flags or other navigational aids or markers shall not be constructed or placed in the Han River Estuary Area by either side except with the approval of the Military Armistice Commission.
9. The Commanders of the opposing sides shall prescribe rules of vessel registration which shall apply to their respective sides. Reports on all vessels registered shall be submitted to and deposited with the Military Armistice Commission for record purposes.
10. Each ship, vessel and water borne craft within the Han River

5. 각방 사령관은 자기측 인원의 민사 행정과 구제 사업을 책임지고 취급한다. 질서 유지와 본 규칙의 각항 규정을 집행하기 위하여 각방은 그 수요에 따라 한강 하구 수역내에 네(4) 척을 넘지 않는 민사 행정 경찰용 순찰 선박과 이십사(24) 명을 넘지 않는 민사 행정 경찰을 제공한다. 민사 행정 경찰의 무기는 권총과 보총에 한한다. 민사 행정 경찰은 자기측의 일체 위반자를 체포하며 자기측의 파손된 선박이 자기측 강안에 도달하도록 협조한다.

6. 민간에서 오랫동안 관습적으로 사용하여 온 한강 하구 수역내에 성문화되지 않은 항행 규칙과 습관은 정전협정중의 각항 규정과 본 규칙에 저촉되는 것을 제외하고는 쌍방 선박이 이를 존중한다.

7. 군사정전위원회의 특정한 허가 없이는 모든 군용 선박과 군사 인원 및 무기, 탄약을 실은 민용선박과 중립국 선박은 모두 한강 하구 수역에 들어가지 못한다.

8. 군사정전위원회의 비준이 없이는 어느 일방이든지 한강 하구 수역내에 부표, 부유물, 등광, 표판, 깃발 기타 항행 보조물 또는 표식물을 설치하지 못한다.

9. 적대 쌍방 사령관은 자기측의 선박 등록에 적용할 규칙을 규정한다. 이미 등록된 모든 선박에 관한 보고는 군사정전위원회에 제출하여 비치케 한다.

10. 한강 하구 수역내에 매개 선박과 수상에서 항행하는

Estuary Area shall be subject to and comply with the rules set forth below:

a. Each ship, vessel or craft shall carry a Registration Card which shall include type of vessel, length and tonnage, nationality of craft, name, and nationality of owner and home port.

b. Each ship, vessel or craft shall be subject to challenge, search and interrogation by personnel of the Joint Observer Teams and the Military Armistice Commission, and by the civil police of its own side.

c. Upon being challenged, each ship, vessel or craft shall furnish the following information:

(1) Nationality of craft and owner

(2) Name of owner

(3) Home port

(4) Port of Departure

(5) Port of Destination

(6) Names of Captain and Crew

(7) Names of Passengers

(8) Type and quantity of Cargo

d. Each ship, vessel, or craft shall prominently display its national flag, standard, or ensign at all times.

e. A civil ship, vessel, or craft shall not mount any military equipment except as authorized by the Military Armistice Commission.

f. A ship, vessel, or craft from one side shall not have access to the waters or shore controlled by the other side, and shall not go closer than one hundred (100) meters to the boundary of the Han River Estuary Area on the other side.

g. A ship, vessel, or craft from one side shall not contact or

교통기재는 하기 규정을 준수하며 복종한다.

ㄱ. 매개 선박은 선박의 형, 길이와 톤수, 선박의 국적, 선주의 성명 및 국적과 선박 등록 항구를 명기한 등록증을 휴대한다.
ㄴ. 매개 선박은 군사정전위원회 및 공동감시소조 인원과 자기측 민사행정 경찰의 조사와 수색과 문의에 복종한다.

ㄷ. 매개 선박은 조사받을 때 하기의 재료를 제공한다.
(1) 선박 및 선주의 국적
(2) 선주의 성명
(3) 선박 등록 항구
(4) 출발항
(5) 목적항
(6) 선장과 선원의 성명
(7) 승객의 성명
(8) 적재 화물의 종류와 수량
ㄹ. 매개 선박은 언제나 자기 국기 또는 국적을 표시하는 깃발을 뚜렷하게 단다.
ㅁ. 어떠한 민용선박이든지 군사정전위원회의 허가 없이는 어떠한 군사장비도 설치하지 못한다.

ㅂ. 일방의 선박은 타방의 통제 수역과 강안에 들어가지 못하며 한강 하구 수역의 타방의 경계선으로부터 백(100) 미터 이내에 접근하지 못한다.

ㅅ. 일방의 선박은 충돌을 피하기 위한 항행 신호를

communicate with a ship, vessel, or craft of the opposite side, except that navigational signals to avoid collision shall be permitted.

h. A ship, vessel, or craft of one side shall not transfer or exchange cargo, equipment, or passengers with a ship, vessel, craft or person of the opposite side.

i. A ship, vessel, or craft shall not sail or operate during the hours of darkness, but shall anchor near the shore of its respective side during the period from one half of an hour after sunset until one half of an hour before sunrise.

11. Personnel of one side shall not have access to the waters or the shore controlled by the other side.

12. Personnel of one side shall not contact or communicate with personnel, ships, vessels and craft of the other side except as authorized by the Military Armistice Commission.

13. When a ship, vessel, or craft sailing in the Han River Estuary Area suffers calamities as a result of a storm, the influence of currents or other catastrophies, both sides have the responsibility to come to its rescue irrespective of the side to which the vessel and personnel may belong. The disposition after the rescue will be left in the charge of the Joint Observer Team.

14. Collisions occurring in the Han River Estuary Area involving ships and personnel of only one side shall be settled in accordance with the laws of that side, but whenever they affect or damage the ships and personnel of the opposite side the Joint Observer Team shall investigate and report the findings to the Military Armistice Commission for such action as agreed upon by that body.

15. Amendments and additions to these Rules may be adopted upon agreement by the Senior Members of both sides on the Military Armistice Commission.

하는 것을 제외하고는 상대방의 선박과 연락이나 통신을 하지 못한다.

ㅇ. 일방의 선박은 상대방의 선박이나 인원과 화물, 장비 또는 승객을 양도하거나 교환하지 못한다.

ㅈ. 어떠한 선박이든지 야간에는 항행이나 활동을 하지 못하며 일몰 반시간 후부터 일출 반시간 전까지의 기간에는 자기측 강안 부근에 정박한다.

11. 일방의 인원은 타방의 통제수역이나 강안에 들어가지 못한다.
12. 일방의 인원은 군사정전위원회의 비준이 없이는 타방의 인원 및 선박과 연락이나 통신을 하지 못한다.

13. 한강 하구 수역내에서 항행하는 선박이 폭풍이나 조류의 영향을 받거나 또는 기타 재해로 인하여 재난을 당하였을 때 그 선박과 인원이 어느 측에 속하였던지를 막론하고 쌍방은 모두 이를 구제할 책임을 지되 구제한 후의 처리는 공동감시소조가 이를 책임진다.

14. 한강 하구 수역내에서 발생한 선박의 충돌 사건이 오직 일방의 선박과 인원에만 관계되는 경우에는 해당측 법률에 의하여 해결한다. 상대방의 선박과 인원에게 위험이나 파손을 입게 한 경우에는 공동감시소조가 조사하여 조사 결과를 군사정전위원회에 보고하며 동 기관은 합의된 행동을 한다.
15. 군사정전위원회 쌍방 수석위원의 협의로써 본 규칙을 수정하며 보충할 수 있다.

16. After these Rules have been ratified by the Military Armistice Commission, the Commanders of the opposing sides shall give them wide publicity. These Rules shall become effective on 10 October 1953.

9. REVISED GENERAL RULES FOR THE CONTROL, ORGANIZATION, OPERATION AND SUPPORT OF THE JOINT OBSERVER TEAMS

1. CONTROL

a. Operational control of the Joint Observer Team shall be exercised by the Military Armistice Commission through the Secretaries appointed by both sides.

b. The Military Armistice Commission, or the Senior Member of each side, shall dispatch Joint Observer Teams through the Secretaries appointed by both sides.

2. ORGANIZATION

a. Five(5) Joint Observer Teams shall be established and the teams separately numbered from ONE to FIVE.

b. Each side shall furnish, to each of such teams, two (2) or three (3) field grade officers and not more than thirty (30) necessary staff assistants and working personnel.

c. Technical personnel necessary to the operation of each Joint Observer Team, such as explosive ordnance disposal personnel, ammunition experts, surveying personnel, photographers, drivers, etc., shall be provided respectively by each side within the total number of staff assistants and working personnel as provided for in

16. 군사정전위원회가 본 규칙을 비준한 후 적대 쌍방 사령관은 이를 광범히 선포하며 1953년 10월 10일부터 효력을 발생한다.*

- 1953년 10월 3일에 군사정전위원회 제22차 회의에서 비준되었음.
- Approved at the 22nd Meeting of the Military Armistice Commission held on 3 October 1953.

9. 공동감시소조의 관리, 조직, 사업 및 보급에 관한 수정 총칙

1. 관리

ㄱ. 군사정전위원회는 쌍방이 임명한 비서장을 통하여 공동감시소조의 사업을 관리한다.

ㄴ. 군사정전위원회 혹은 각방 수석위원은 쌍방이 임명한 비서장을 통하여 공동감시소조를 파견한다.

2. 조직

ㄱ. 오(5) 개의 공동감시소조를 조직하고 제 일(1)부터 제 오(5)까지의 번호를 매긴다.
ㄴ. 각방은 이러한 매개 소조에 이(2)명 내지 삼(3)명의 영급 장교와 삼십(30)명을 초과하지 않는 필요한 참모 보조인원 및 공작 인원을 제공한다.
ㄷ. 폭파물 처리 인원, 탄약 전문가, 측량 인원, 촬영 인원, 운전수 등 각 공동감시소조의 사업에 필요한 기술 인원은 상기 항목에 규정한 참모 보조인원과 공작 인원 총수내에서 각방이 각기 제공한다.

the above paragraph.

3. DEPLOYMENT

a. The Han River Estuary shall constitute area number ONE. The Demilitarized Zone shall be divided into four (4) areas, numbered from TWO through FIVE. Each Joint Observer Team shall perform the mission as provided for in Paragraph 26 of the Armistice Agreement in the correspondingly numbered area.

b. Upon the suggestion of the Secretary of either side, boundaries of the numbered area, both in the Han River Estuary and in the Demilitarized Zone, May be adjusted through negotiation as accessibility, terrain features and other considerations of operational conveniences may require.

c. The two components of the Joint Observer Teams shall billet in the territory under the military control of their respective sides, but at locations from which they can readily reach their respective areas in the Han River Estuary or the Demilitarized Zone. The teams may be dispatched from the billeting locations to the specific areas of their assignment or to any other areas in the Han River Estuary or the Demilitarized Zone. Teams billeted in or adjacent to assigned areas are not to be considered as dispatched.

4. METHODS AND PROCEDURE OF OPERATION

a. The functions of the Teams shall be performed jointly by the team members of both sides.

b. Teams shall meet at times and places mutually agreed upon by the senior team members of both sides, or at times and places directed by the Military Armistice Commission.

c. Teams are not to be considered as dispatched when they perform

3. 분포

ㄱ. 한강 하구는 일(1) 개 지역으로 하여 번호는 제 일(1)로 한다. 비무장지대는 사(4) 개 지역으로 나누어 번호는 제 이(2)부터 제 오(5)까지로 한다. 각 공동감시소조는 그 번호에 해당한 지역내에서 정전협정 제26항에 규정한 임무를 집행한다.

ㄴ. 한강 하구 및 비무장지대내의 각 번호 지역의 경계선은 출입의 편리, 지형물 및 사업상 편리에 근거한 고려에 의하여 어느 일방의 비서장이 제출한 후에 이를 협의 조정한다.

ㄷ. 공동감시소조의 쌍방 성원은 자기측의 군사 통제 지역내에 각각 주재하되 그 주재 지점은 한강 하구 또는 비무장지대내의 자기측 지역에 신속히 도달할 수 있는 곳으로 한다. 각 소조는 그가 주재하는 지역으로부터 그에게 지정된 특정한 지역 또는 한강하구나 비무장지대의 임의의 기타 지역에 파견될 수 있다. 지정된 지역내에 또는 이에 인접하여 주재하는 각 소조는 파견된 것으로 인정하지 않는다.

4. 사업 방법과 절차

ㄱ. 소조의 각종 직책은 소조의 쌍방 조원이 공동으로 이를 집행한다.

ㄴ. 각 소조는 쌍방 수석 조원이 호상 동의한 시간과 지점에서 회합하거나 또는 군사정전위원회가 지시한 시간과 지점에서 회합한다.

ㄷ. 각 소조가 일상적 임무를 집행할 때에는 이를 파

routine duties.

d. While investigating reported violations of any provisions of the Armistice Agreement pertaining to the Demilitarized Zone and the Han River Estuary, team members of the two sides shall have equal opportunity to collect facts and listen to the testimonies of the witnesses.

5. REPORTS

a. Reports shall be made after a Joint Observer Team has been dispatched to investigate violations of the Armistice Agreement, in accordance with Paragraph 27 of the Armistice Agreement, after directed to participation in any joint activity, or when deemed necessary by the individual teams. Reports shall be made in the appropriate forms indicated in Annex 1 and Annex 2.

b. Reports of a Joint Observer Team shall be made in accordance with the following procedures:

(1) Prior to the preparation of a report, personnel from the two components of the Team shall meet and, through discussion, ascertain fully and clearly all the facts and points upon which the two components are in agreement and those with respect to which disagreement exists.

(2) Each of the components shall then separately draft, in the one of the three(3) official languages with which it is most familiar, its report in which it shall set forth:

(a) All the facts and points upon which agreement has been reached.

(b) Two component's view upon the facts and points not agreed upon.

(3) Each component shall next transmit promptly to the Military Armistice Commission, by telephone, radio, telegraph or other

견된 것으로 인정하지 않는다.

ㄹ. 정전협정중의 비무장지대와 한강 하구에 관한 어떠한 규정의 위반을 보고에 의하여 조사할 때 소조의 쌍방 조원은 사실을 수집하며 증인의 증언을 청취할 동등한 기회를 가진다.

5. 보고

ㄱ. 공동감시소조는 매번 정전협정 제 27항에 따라 파견되어 정전협정 위반사건을 조사한 후 또는 명령에 의하여 어떠한 공동 활동에 참가한 후 또는 각 개별적 소조가 필요하다고 인정할 때 보고를 제출한다. 보고는 부록 1과 부록 2에 표시된 해당 격식에 의하여 작성한다.

ㄴ. 공동감시소조의 보고서는 하기 절차에 의하여 작성한다.

(1) 보고서를 작성하기 전에 소조의 쌍방 성원은 회합하여 토의를 거쳐 소조의 쌍방 성원이 합의를 본 모든 사실과 점들 및 의견 차이가 있는 모든 사실과 검들을 충분히 또 명백히 확정한다.

(2) 각방 성원은 삼(3) 개 공용어 중 자기가 가장 능통한 한 가지 글로 보고서를 각기 작성하되 이 보고서에는 하기사항을 기술한다.

(ㄱ) 합의에 도달한 모든 사실과 점들.

(ㄴ) 합의를 보지 못한 사실과 점들에 대한 자기측 성원의 의견.

(3) 그 다음 각방 성원은 전화, 라디오, 전신 또는 기타 신속한 통신 기재로 그 보고서의 주요 내용을 군사정

means of rapid communication, the material contents of its report.

(4) Reports of each component will be prepared in four(4) copies, each of which will be signed by its senior member and distributed as follows:

Two(2) forwarded promptly to the Military Armistice Commission, one(1) delivered to the other component of the Team and one(1) retained in the files of the originating component.

(5) Each component shall also retain in its files the one(1) copy received of the report of the other component upon the same subject.

6. LOGISTIC SUPPORT

a. Members of both sides on a Joint Observer Team shall be logistically supported by their own side at their billeting locations.

b. In addition to billeting and logistically supporting his members of the Joint Observer Teams, each Commander shall furnish all equipment and supplies required by them, including transportation means and communication services.

7. AMENDMENT

Provisions herein, if found to be inappropriate after tentative application, may, upon the suggestion of the Secretary of either side, be amended through negotiation between the Secretaries of the two sides.

전위원회에 신속히 전달한다.

(4) 각방 성원은 보고서를 네(4) 벌 작성하고 매벌에 자기측 수석조원이 서명하여 다음과 같이 배부한다. 두(2) 벌은 군사정전위원회에 신속히 송부하며, 한 벌은 동 소조의 상대방 성원에게 교부하며, 한 벌은 보고서를 작성한 자기측 성원의 문건철에 보관한다.

(5) 각방 성원은 또한 상대방 성원으로부터 받은 같은 건에 관한 보고서를 자기측 문건철에 보관한다.

6. 보급

ㄱ. 공동감시소조의 쌍방 조원은 그의 주재지에서 자기측으로부터 보급을 받는다.

ㄴ. 각방 사령관은 공동감시소조의 자기측 조원에게 숙소와 보급을 공급하는 외에 그에 필요한 운수 기재와 통신 복무를 포함한 일체 장비와 공급을 제공한다.

7. 수정

본 총칙의 각항 규정을 잠정적으로 집행하는 과정에서 부적당한 점을 발견할 때에는 어느 일방의 비서장이 제출한 후 쌍방 비서장이 합의하여 이를 수정할 수 있다.

• 본 수정 총칙은 1955년 7월 18일의 제135차 비서장 회의에서 국련군측과 조중측에 의하여 합의를 보았으며, 1955년 7월 26일 공동일직장교들을 통해서 동 총칙의 비준된 언어 문본이 교환되었음.

• Revised rules agreed to by UNC and KPA/CPV at 135th Secretaries' meeting, 18 July 1955, and approved language versions exchanged via Joint Duty Officers, 26 July 1955.

MILITARY ARMISTICE COMMISSION

Date: ____________

Place: ____________

Ser No: ___________

From: ________________________ Component

Joint Observer Team Number ________________

To: Military Armistice Commission

Via: Secretaries, Military Armistice Commission

Subject: Report on _________________________________ (Joint Activity)

1. The following report is submitted on ________________________:

(So many of the following items, and other actions, as may apply to the activity concerned:

a. Meeting of the Teams

b. Arrangements by the Teams

c. Implementation of the Program.

d. Results.

e. Recommendations and conclusions.)

____________________ (Signature)

____________________ (Name)

____________________ (Rank)

Senior Member

(Component) Annex 1 to TAB "H" (1)

군사정전위원회

날　짜 ______________
장　소 ______________
순번호 ______________

발신자:　공동감시소조 제조

______________ 성원

수신자:　군사정전위원회

경유자:　군사정전위원회 쌍방 비서장

건　명:　______________________________ 에 관한 보고.

1. ____ 에 관한 보고를 하기와 같이 제출한다.

(하기 각항 및 관계 활동에 적용될 기타 행동을 기입한다.

ㄱ. 소조의 회합

ㄴ. 소조가 취한 조치

ㄷ. 사업의 실시

ㄹ. 결과

ㅁ. 건의와 결론)

____

(성원)

수석 조원

____

(계급)

____

(성원)

____

(서명)

색인표 H(1)에 대한 부록 1

MILITARY ARMISTICE COMMISSION

Date: ____________

Place: ____________

Ser No: ____________

From: ____________________ Component

Joint Observer Team Number ________

To: Military Armistice Commission

Via: Secretaries, Military Armistice Commission

Subject: Report of Investigation of Alleged Violation Reference: (a)

1. Pursuant to reference(a), an investigation was conducted and the following report thereon is submitted:

a. Summary of evidence with attached statements of witnesses, etc. b. Findings: c. Recommendations:

______________ (Signature)

______________ (Name)

______________ (Rank)

Senior Member

______________ (Component)

(list enclosures and exhibits as attached)

Annex 2 to TAB "H" (1)

군사정전위원회

날 짜 ______________

장 소 ______________

순번호 ______________

발신자: 공동감시소조 제조

______________ 성원

수신자: 군사정전위원회

경유자: 군사정전위원회 쌍방 비서장

건 명: 보고된 위반 사건 조사에 관한 특별 보고.

접수 문건: (가)

1. 접수 문건 (가)에 따라 조사를 진행하고 그에 대하여 다음과 같은 보고서를 제출한다.

ㄱ. 증인들의 증언을 첨부한 증거의 종합 진술 기타

ㄴ. 판정서:

ㄷ. 건의:

____

(성원)

수석 조원

____

(계급)

____

(성명)

____

(서명)

(첨부하는 유첨물 및 전시물들을 열거한다)

색인표 H(1)에 대한 부록 2

10. CHRONOLOGY OF NEGOTIATION FOR REDUCTION OF JOINT OBSERVER TEAMS AND REVISED GENERAL RULES FOR JOINT OBSERVER TEAMS

REFERENCES;

The original Tentative General Rules for the JOTs providing for 10 teams was approved at the 15th MAC Meeting, 28 August 1953.

First Amendment to the General Rules for JOTs was approved at 93rd Secretaries' Meeting, 30 Nov 53. The amendment specified the manning of JOT #1 and #2 as Han River Estuary teams.

The Second Amendment to the General Rules for JOTs was approved at the 98th Secretaries' Meeting, 30 Dec 53. This amendment eliminated the weekly reports required of each JOT.

UNC proposed at the 35th MAC Meeting, 10 Jan 54 the reduction of JOTs from 10 to 7 (KPA/CPV agree in principle.) UNC submits written plan for reduction at 102nd Secretaries' Meeting, 22 Jan 54.

KPA/CPV make counter-proposal of reduction of JOTs from 10 to 6 at 103rd Secretaries' Meeting, 29 Jan 54. (UNC accepts plan for study).

UNC agrees to KPA/CPV counter-proposal for reduction of JOTs from 10 to 6 at 104th Secretaries' Meeting, 6 Feb 54.

UNC submits at 105th Secretaries' Meeting, 19 Feb 54, Revised General Rules for JOTs incorporating reduction 10 to 6. KPA/CPV accepts for study.

10. 공동감시소조의 숙소 및 수정 총칙을 위한 협의 연대표

참조: 10개 소조를 규정하고 있는 공동감시소조에 관한 원래의 잠정적 총칙은 1953년 8월 28일의 제15차 군사정전위원회 회의에서 비준되었다. 공동감시소조 총칙에 대한 최초의 수정안은 1953년 11월 30일의 제93차 비서장 회의에서 비준되었으며 해 수정안은 공동감시소조 제1조 및 제2조를 한강 하구 담당 소조로 할 것을 명확히 규정하였다. 공동감시소조 총칙의 두 번째 수정안은 1953년 12월 30일의 제98차 비서장회의에서 비준되었으며 해 수정안은 매개 공동감시소조가 주간 보고서를 제출하는 것을 폐기시켰다.

1954년 1월 10일의 제35차 군사정전위원회 회의에서 국제연합군측은 공동감시소조를 10개 소조로부터 7개 소조로 축소할 것을 제의하였다. (조중측은 원칙적으로 이에 동의하였다.) 국제연합군측은 1954년 1월 22일의 제102차 비서장 회의에서 축소 계획서를 제출하였다.

조중측은 1954년 1월 29일 제103차 비서장 회의에서 공동감시소조를 10개 소조로부터 6개 소조로 축소하자는 반대 제안을 하였다. (국제연합군측은 검토하기 위하여 그 제안을 접수하였다.)

국제연합군측은 1954년 2월 6일 제104차 비서장 회의에서 공동감시소조를 10개 소조로부터 6개 소조로 축소하자는 조중측의 반대 제안에 동의하였다.

국제연합군측은 1954년 2월 19일의 제105차 비서장 회의에서 10개 소조로부터 6개 소조로 축소 통합하는 공동감시소조의 수정 총칙안을 제출하였으며 조중측은 검토

Secretaries of both sides formally approved Revised General Rules for JOTs incorporating reduction of JOTs 10 to 6 at 106th Secretaries' Meeting, 19 Mar 1954.

UNC proposes at 129th Secretaries' Meeting, 8 Apr 55 to reduce JOTs from 6 to 4. (KPA/CPV agrees in principle).

KPA/CPV makes counter-proposal at 134th Secretaries' Meeting, 9 July 1955 for reduction of JOTs from 6 to 5.

UNC agrees at 135th Secretaries' Meeting to KPA/CPV proposal for reduction of JOTs from 6 to 5 and submits revised rules incorporating reduction from 6 to 5 (KPA/CPV accepts revised rules for study).

KPA/CPV and UNC exchange jointly approved language versions for "Plan for Reduction of JOTs from 6 to 5" and "Revised Rules for JOTs" at JDO Meeting, 26 July 1955. (KPA/CPV also confirms approval of Revised Rules and Reduction via message through JDO, 26 July 1955).

OMAR C. ESTES
Major, USAF
Ch, Admin Br, Sct

하기 위하여 이를 접수하였다.

쌍방 비서장은 1954년 3월 19일 제106차 비서장 회의에서 공동감시소조를 10개 소조로부터 6개 소조로 축소 통합하는 공동감시소조의 수정 총칙을 정식으로 비준하였다.

국제연합군측은 1955년 4월 8일의 제129차 비서장 회의에서 공동감시소조를 6개 소조로부터 4개 소조로 축소할 것을 제의하였다. (조중측은 원칙적으로 이에 동의하였다.)

조중측은 1955년 7월 9일 제134차 비서장 회의에서 공동감시소조를 6개 소조로부터 5개 소조로 축소하자는 반대 제안을 하였다.

국제연합군측은 제135차 비서장 회의에서 공동감시소조를 6개 소조로부터 5개 소조로 축소하자는 조중측 제의에 동의하였으며 6개 소조로부터 5개 소조로 축소 통합하는 수정 총칙을 제출하였다. (조중측은 동 · 수정 총칙을 검토하기 위하여 접수하였다.)

조중측과 국제연합군측은 1955년 7월 26일의 공동일직장교 회의에서 "공동감시소조를 6개 소조로부터 5개 소조로 축소하는 계획서"와 "공동감시소조의 수정 총칙"의 공동으로 비준된 각 해당 언어의 문본은 교환하였다. (조중측은 1955년 7월 26일의 공동일직장교를 통하여 교환된 통지문에 의하여 수정총칙과 축소안의 비준을 확인하였다.)

비서처 행정과장
미국 공군 소령
오마 씨. 에스테스

PLAN FOR REDUCTION OF JOINT OBSERVER TEAMS FROM SIX TO FIVE

JOINT OBSERVER TEAM NUMBER 1

The Han River Estuary as presently defined,

The Team meeting place - as present.

JOINT OBSERVER TEAM NUMBER 2

From: Western limit of DZ at MDL marker 0001.

To: Marker 0343 inclusive.

This includes the entire area of the present Team 2 and the part of the present 3 that is west of the stream flowing from Hoesan-dong to the Imjin-Gang and west of the Imjin-Gang.

The Team meeting place - meeting place of the present team Number 2.

JOINT OBSERVER TEAM NUMBER 3

From: Eastern limit of Team 2

To: Marker 0710 inclusive.

This includes the part of the present Team 3 that is east of the stream flowing from Hoesan-dong to the Imjin-Gang and east of the eastern edge of the ImjinGang, and the part of the present Team 4 that is west of the Namdae-chon.

The Team meeting place - meeting places of the present Team 3 and Team 4, to use either of them according to the Team's requirement.

JOINT OBSERVER TEAM NUMBER 4

From: Eastern limit of Team 3

To: Marker 0982 inclusive.

공동감시소조를 6개 소조로부터 5개 소조로 축소하는 데 관한 계획

공동감시소조 제1조

한강하구는 현재 규정되어 있는 바와 같음.

소조의 회의 장소 · 현재와 동일함.

공동감시소조 제2조

비무장지대 최서단 군사분계선 표식물 제0001호로부터 표식물 제0343호를 포함하는 지점까지.

이는 현재의 제2조 담당구역 전부 및 제3조 담당구역 중 혜산동으로부터 임진강으로 흐르는 개천의 서쪽과 임진강의 서쪽 부분을 포함한다.

소조의 회의 장소 · 현재 제 2조의 회의 장소.

공동감시소조 제3조

제2조의 동단부터 표식물 제0710호를 포함하는 지점까지.

이는 현재의 제3조 담당구역중 혜산동으로부터 임진강으로 흐르는 개천의 동쪽 부분과 임진강 동단부터 동쪽 부분 및 현재의 제4 공동감시소조 담당구역중 남대천의 서쪽 부분을 포함한다. 소조의 회의장소 · 소조의 필요에 따라 현재의 제3조와 제4조의 회합 장소중 어느 하나를 사용한다.

공동감시소조 제4조

제 3조의 동단부터 표식물 제0982호를 포함하는 지점까지.

This includes the part of the present Team 4 that is east of the eastern edge of the Namdae-chon and the part of the present Team 5 that is west of the Suip-chon.

The Team meeting place - on the MDL in the vicinity of Korisil (38° 19' 12" North, 127° 40' 16" East).

JOINT OBSERVER TEAM NUNMBER 5

From: Eastern limit of Team 4.

To: Eastern limit of DZ at marker 1292.

This includes the part of the present Team 5 that is east of the eastern edge of the Suip-chon and the entire area of the present Team 6.

The Team meeting place - on the MDL in the vicinity of Yangjimal (38° 18' 50" North, 127° 59' 24" East) and the meeting place of the present Team 6, to use either of them according to the Team's requirement.

이는 현재의 제4조 담당구역중 남대천의 동단부 동쪽 부분 및 현재의 제5조 담당구역중 수입천의 서쪽 부분을 포함한다.

소조의 회의 장소 · 고리실 부근에 있는 군사분계선상(북위 38도 19분 12초 동경 127도 40부 16초).

공동감시소조 제5조

제4조의 동단부터 비무장지대의 최동단 표식물 제1292호까지.

이는 현재의 제5조 담당구역중 수입천 동단부의 동쪽 부분 및 현재의 제6조 담당구역 전부를 포함한다.

소조의 회의 장소 · 양지말 부근의 군사분계선상(북위 38도 18분 50초, 동경 127도 59분 24초)와 현재의 제6조의 회의 장소중 등 소조의 필요에 따라 어느 하나를 사용한다.

- 해 공동감시소조 축소 계획은 1955년 7월 18일의 제135차 비서장 회의에서 비준되었으며 1955년 7월 26일 공동일직장교를 통하여 비준된 해당 언어 문본이 교환되었음.
- Plan for Reduction of JOTs approved at 135th Secretaries' meeting, 18 July 1955, and approved language version exchanged via Joint Duty Officers, 26 July 1955.

17. PROCEDURES FOR SUPERVISING AND REPORTING THE ARRIVALS IN AND DEPARTURES FROM KOREA OF MILITARY PERSONNEL AND REPLACEMENT ITEMS.

(1) Written reports concerning arrivals in and departures from Korea of rotation personnel of each side shall be submitted daily to the Military Armistice Commission and the Neutral Nations Supervisory Commission within seventy-two (72) hours after arrivals and departures; such reports shall include places of arrival and departure and number of persons arriving at or departing from each such place. For the convenience of checking and computation, cumulative totals of rotation personnel arriving in and departing from Korea shall be reported daily. The Neutral Nations Supervisory Commission, through its Neutral Nations Inspection Teams, shall conduct supervision and inspection of rotation personnel at the specified ports of entry.

(2) Written reports concerning the introduction into and departure from Korea of persons of each side on temporary duty or for short periods of leave outside Korea shall be made daily by each side to the Military Armistice Commission and the Neutral Nations Supervisory Commission within seventy-two (72) hours after such arrivals and departures. Such reports shall include places of arrival and departure and the number of persons arriving at or departing from each such place. For the convenience of checking and computation, cumulative totals of such persons arriving in and departing from Korea on temporary duty or for short periods of leave outside Korea shall be reported daily.

(3) Reports concerning the introduction into and the removal from Korea of every shipment of combat aircraft, armored vehicle,

17. 군사 인원의 한국에의 도착과 한국으로부터의 이거와 물건의 교체를 감독하며 보고하는 데 관한 절차

(1) 각방의 윤환 인원의 한국에의 도착과 한국으로부터의 이거는 도착과 이거 칠십이(72) 시간 내에 매일 군사정전위원회 및 중립국감독위원회에 서면 보고를 제출한다. 이 보고는 입경과 출경의 지점 및 매개지점에서 입경하는 인원과 출경하는 인원의 숫자를 포함한다. 검사 대조와 계산에 편리하게 하기 위하여 매일 한국에 도착하며 한국으로부터 이거하는 윤환 인원의 누계 총수를 보고 한다. 중립국감독위원회는 중립국시찰소조를 통하여 특정한 출입항에서 윤환 인원을 감독하며 시찰한다.

(2) 임시 임무를 담당하거나 혹은 한국 경외에서 단기 휴가를 하는 각방의 인원은 도착과 이거 칠십이(72) 시간내에 매일 군사정전위원회 및 중립국감독위원회에 서면 보고를 제출한다. 이 보고는 입경과 출경의 지점 및 매개지점에서 입경하는 인원과 출경하는 인원의 숫자를 포함한다. 검사 대조와 계산에 편리하게 하기 위하여 매일 한국에 도착하며 한국으로부터 이거하는 임시 임무를 담당하는 이러한 인원 혹은 한국 경외에서 단기 휴가를 하는 이러한 인원의 누계 총수를 보고 한다.

(3) 각방이 교체하는 작전 비행기, 장갑차량, 무기 및 탄약의 한국에의 반입과 한국으로부터의 반출은 매차

weapons, and ammunition by each side for replacement purposes shall be submitted daily to the Military Armistice Commission and the Neutral Nations Supervisory Commission not later than ten(10) days after the introduction into or the removal from Korea. Such reports shall include statements regarding the disposition of the items being replaced. To afford the Neutral Nations Inspection Teams adequate opportunity for inspection and supervision, local port authorities shall provide the Neutral Nations Inspection Teams with advance information on all shipments.

반입과 반출 후 늦어도 십(10) 일내에 매일 이를 군사정전위원회 및 중립국감독위원회에 보고한다. 이 보고중에서 교체되는 물건의 처리 정형을 설명한다. 당지 출입항 당국은 중립국 시찰소조에 매차의 재료를 사전에 제공하여 동 소조로 하여금 시찰하며 감독하는 적당한 기회를 가지게 한다.

- 1953년 9월 28일에 열린 군사정전위원회 제21차 회의에서 비준되었음.
- Approved at the 21st Meeting of the Military Armistice Commission held on 28 September 1953.

21. Agreement on reduction of NNITS (CHRONOLOGY OF NEGOTIATION FOR REDUCTION AND REDESIGNATION OF NNITS)

REFERENCES:

1. At 189th Plenary session of NNSC. 13 April 1955, Swedish delegation submitted a proposal to withdraw all NNITS from ports. Proposal put on agenda for next meeting.

2. AT 190th Plenary session of NNSC, 15 April 1955, Czech delegation asked for discussion on proposal be postponed until next meeting.

3. At 191st Plenary session of NNSC, 20 April 1955, Czech and Polish delegation submitted a counter proposal to reduce NNITS at all ports so that teams would be composed of two nationalities instead of four, thereby reducing personnel required by 50%.

4. At 193rd Plenary session of NNSC, 3 May 1955, unanimous agreement was reached and the following letter to the MAC was drafted and signed by all four delegations:

"NEUTRAL NATIONS SUPERVISORY COMMISSION

Panmunjom, 3 May 1955
From: Neutral Nations Supervisory Commission
To: Military Armistice Commission

21. 중립국 시찰 소조의 축소와 재임명에 관한 협의

참조:

1. 1955년 4월 13일의 중립국감독위원회 제189차 본회의에서 서전(스웨덴) 대표는 각 출입항으로부터 모든 중립국 시찰 소조를 철수시키자는 제의를 제출하였다. 해 제의는 다음 회의의 의제로 채택되었다.
2. 1955년 4월 15일의 중립국감독위원회 제190차 본회의에서 체코슬로바키아 대표는 해 제의의 토의를 다음 회의까지 연기할 것을 제의하였다.
3. 1955년 4월 20일의 중립국감독위원회 제191차 본회의에서 체코슬로바키아와 파란(폴란드) 대표들은 모든 출입항의 중립국 시찰 소조를 4개국 대신에 2개국으로 구성함으로써 필요한 인원을 50% 감소하자는 반대 제안을 제출하였다.
4. 1955년 5월 3일의 중립국감독위원회 제193차 본회의에서 만장일치의 합의를 보았으며 군사정전위원회에 보내는 하기의 서한이 4개국 대표 전원에 의하여 기초, 서명되었다.

중립국감독위원회

1955년 5월 3일 판문점
발신자: 중립국감독위원회
수신자: 군사정전위원회

The NNSC has its 193rd Plenary Meeting following the Swedish and Swiss proposals unanimously decided to recommend to the MAC to agree to the following changes regarding the NNITS:

1. The NNITS No. 2 (Changjin), No. 3 (Hungnam), No. 9 (Kangnung) No. 10 (Kunsan) to be withdrawn for the time being; it being understood that this measure does not modify the status of the said Ports of Entry according to the Armistice Agreement.
2. The remaining NNITS to be composed as subteams i.e. by not less than two(2) members, one of whom shall be from a Neutral Nation nominated by the Commander-in-Chief United Nations Command, and one of whom shall be from Neutral Nation nominated jointly by the Supreme Commander of the Korean People's Army and the Commander of the Chinese People's Volunteers.

The above-mentioned changes should become effective a week after the NNSC has been informed of the agreement of the MAC.

For details concerning the discussion of the above-mentioned matter, see NNSC records No. 189, 191, and 193 dated April 13, 20 and May 3 respectively.

/s/ Col W. Zema
Acting Czechoslovak member
/s/ W. Tykocinski
Polish Member

/s/ T. Greenwall
Swedish Member
/s/ K. Stucki
Swiss Member

5. On 4 May 1955 General Carter acknowledged letter and stated UNC would consider, and reply to letter at a later date.

중립국감독위원회는 제193차 본회의에서 서전과 서서(스위스)의 제의에 따라 중립국 시찰 소조에 대한 다음과 같은 변경에 동의하도록 군사정전위원회에 건의할 것을 만장일치로 결정하였다.

1. 제2(청진) 제3(홍남) 제9(강릉) 제10(군산) 중립국 시찰 소조들이 당분간 철수할 것이며 이 조치는 정전협정에 의한 상기 출입항들의 자격을 변경하지 않는 것으로 이해될 것이다.
2. 나머지 중립국 시찰 소조들은 분조들로 즉, 최소 2명의 조원들로 구성되며 그 중 1명은 국제연합군 총사령관이 지명한 1개의 중립국으로부터 나오며, 다른 1명은 조선인민군 최고사령관과 중국인민지원군 사령원이 공동으로 지명한 1개 중립국으로부터 나온다.

상기 변경은 중립국감독위원회가 군사정전위원회의 동의에 대하여 통지를 받은 1주일 후부터 효력을 발생한다.

상기 문제에 관한 상세한 토론은 각각 4월 13일, 20일 및 5월 3일부 제189차, 191차 및 제193차 중립국감독위원회 회의록들을 보라.

/서명/ 더불유. 제물랴 대령
체코슬로바키야 위원 대리

/서명/ 티. 그렌발
서전 위원

/서명/ 더불유. 띠꼬린스키
파란 위원

/서명/ 케이. 스툭키
서서 위원

5. 1955년 5월 4일에 카터 장군은 서한을 확인하였으며 국제연합군은 고려할 것과 그 서한에 대한 회신을 후

6. On 23 July 1955, letter from UNC, signed by General Carter stating UNC planned to discontinue the use of the port of Taegu and to use Kunsan instead.
7. At 206th Plenary session of NNSC, 3 Aug 1955, original letter revised, signed, and submitted the MAC to reflect change from Taegu to Kunsan.

"Panmunjom, 3 August 1955"
From: Neutral Nations Supervisory Commission
To: Military Armistice Commission
Subject: Amendment to the recommendation-letter from NNSC to MAC dated 3 May 1955 concerning changes regarding NNITS.

At the 206th Plenary Meeting of 27 July 1955 the Neutral Nations Supervisory Commission has taken note of the letter of July 23, 1955 from Senior Member of UNC, MAC, MAJ, GEN. H.C. Parks, concerning the discontinuation of utilization of the Port of Entry TAEGU and re-utilization of the air facilities at the Port of Entry of KUNSAN.

With reference to its letter sent to the Military Armistice Commission on 3 May 1955 concerning changes regarding NNITS, the Neutral Nations Supervisory Commission has at its 206th and 207th Plenary Meetings agreed to amend the said letter to the effect that in paragraph one(1) the Port of Entry "No. 10 (Kunsan)" be replaced by the Port of Entry "No. 7 (Taegu)".

일에 하겠다고 언명하였다.

6. 1955년 7월 23일부 국제연합군은 대구공항의 사용을 중지하며 대신에 군산을 사용할 것을 계획하고 있다는 카터 장군이 서명한 국제연합군측의 서한.
7. 1955년 8월 3일의 제206차 중립국감독위원회 본회의에서 대구를 군산으로 변경함을 반영하는 서한 원본에 수정을 가하고 서명한 다음 이를 군사정전위원회에 제출하였다. 즉,

발신자: 중립국감독위원회 1955년 8월 3일 판문점
수신자: 군사정전위원회
건 명: 중립국감독위원회로부터 군사정전위원회에 보낸 중립국 시찰 소조에 관계되는 변경에 관한 1955년 5월 3일부 건의 서한에 대한 수정.

1955년 7월 27일의 제206차 본회의에서 중립국감독위원회는 대구 출입항의 사용 중지와 군산 출입항 시설의 재사용에 관한 군사정전위원회 국제연합군측 수석 위원 에이치. 씨. 파크스 소장으로부터의 1955년 7월 23일부 서한에 주의하였다.

1955년 5월 3일 군사정전위원회에 발송된 중립국 시찰 소조에 관계되는 변경에 관한 서한에 관하여 중립국감독위원회는 제206차 및 제207차 본회의에서 첫 구절중 제10 출입항(군산)을 제7 출입항(대구)로 대치한다는 취지로 해 서한 내용을 수정하는 데 동의하였다.

/s/ A. Morski__________	/s/ C. Stucki__________
Minister Morski	Minister Stucki
Polish Member	Swiss Member
/s/ Jens Malling__________	/s/ Taus______________
Ge. Malling	Gen. Taus
Swedish Member	Czechoslovak Member

8. At 66th MAC meeting, 29 August 1955, UNC stated agreement to NNSC proposal and proposed a letter be signed by Senior Members of both sides. KPA/CPV proposed addition of one paragraph:

"2. It is requested that the Military Armistice Commission be informed immediately of said changes."
UNC proposed recess of MAC meeting for a Secretaries Meeting to draft letter.

9. At 138th Secretaries Meeting, 29 August 1955, letter drafted.

10. At 66th MAC Meeting, 29 August 1955 (Reconvened), both sides agreed to following letter which was signed by both Senior members and delivered to NNSC by UNC Secretary immediately after adjournment of meeting.

"August 29, 1955"
To: Neutral Nations Supervisory Commission
From: Military Armistice Commission
1. This is to advise you that the Military Armistice Commission

/서명/ 에이. 모스키 /서명/ 씨. 스룩키
파란 위원 서서 위원

/서명/ 젠스 말링 /서명/ 타우스
서전 위원 체코슬로바키아 위원

8. 1955년 8월 29일의 제66차 군사정전위원회 회의에서 국제연합군은 중립국감독위원회의 제의에 동의하고 쌍방 수석위원이 해 서한에서 서명할 것을 제의하였다. 조선인민군 및 중국인민지원군측은 1개 구절을 다음과 같이 첨가할 것을 제의하였다.
"2. 상기 변경 사항을 즉시 군사정전위원회에 통고할 것을 요망한다."
국제연합군은 당해 군사정전위원회 회의를 휴회하고 비서장 회의를 열어 서한 초안을 작성하게 할 것을 제의하였다.
9. 1955년 8월 29일에 열린 제138차 비서장 회의에서 서한 초안이 작성되었다.
10. 1955년 8월 29일의 제66차 군사정전위원회 회의(재개된)에서 쌍방은 쌍방 수석위원이 서명한 하기 서한에 동의하였으며 해 서한은 당해 군사정전위원회 회의가 휴회된 직후 국제연합군측 비서장에 의하여 중립국감독위원회에 전달되었다.

수신자: 중립국감독위원회 1955년 8월 29일
발신자: 군사정전위원회

1. 군사정전위원회는 중립국감독위원회가 1955년 8월 3

interposes no objection to the Neutral Nations Supervisory Commission proceeding immediately with the implementation of its recommendation contained in its letter of 3 May 1955, as amended by its letter of 3 August 1955.

2. It is requested that the Military Armistice Commission be informed immediately upon completion of the said changes.

/s/ JUNG KOOK ROK
JUNG KOOK ROK
Major General
Korean people's Army
Senior Member
Korean People's Army and
Chinese People's Volunteers Side

/s/ HARLAND C. PARKS
HARLAND C. PARKS
Major General
United States Air Force
Senior Member
United Nations Command Side

UNC copy delivered to NNSC by UNC Secretary on 29 August 1955.

11. Letter from NNSC to MAC, dated 8 September 1955, stated: The recommendations of their letter of 3 May 1955, have been implemented.

12. At the 70th MAC meeting on 31 May 1956, UNC stated: that due to KPA/CPV intransigence in reporting introduction of combat materiel and collaboration of Czech/Pole representatives to block action of NNITs in the north, it was suspending provision of those parts of the Armistice Agreement governing action of NNSC and NNITs in the territory under its control.

일부 중립국감독위원회 서한에 의하여 수정된 바와 같이 1955년 5월 3일부 동 위원회 서한에 표명된 건의안을 즉시 실행하는 데 대하여 아무런 이의가 없다는 것을 통고하는 바이다.

2. 상기 변경이 완료되는 대로 즉시 군사정전위원회에 통고해 줄 것을 요망한다.

/서명/ 정 국 록	/서명/ 할란 씨. 파크스
조선인민군 소장	미합중국 공군 소장
조선인민군 및 중국인민지원군측	국제연합군측
수석위원	수석위원
정 국 록	할란 씨. 파크스

국제연합군측 사본은 1955년 8월 29일에 국제연합군측 비서장에 의하여 중립국감독위원회에 전달되었다.

11. 중립국감독위원회가 군사정전위원회에 보낸 1955년 9월 8일부의 서한은 다음과 같이 지적하였다: 1955년 5월 3일부로 중립국감독위원회가 보낸 서한중의 건의사항들은 실시되었다.

12. 1956년 5월 31일에 진행된 군사정전위원회 제70차 회의에서 국제연합군측은 다음과 같이 발언하였다: 조선인민군 및 중국인민지원군측이 작전 물자의 반입을 보고함에 있어서 비타협적이며 체코슬로바키아와 파란 대표들이 북한에서 중립국 시찰 소조의 활동을 봉쇄하기 위하여 결탁하고 있느니만큼 국제연합군은 정전협정중에서 중립국감독위원회와 중립국 시찰 소

13. Letter dispatched to NNSC from the Senior Member, UNCMAC, dated 31 May 1956, delivered by the JDO, stated: suspension to be put in effect in about one week and requested the withdrawal of teams at that time.

14. Letter from NNSC to MAC, dated 5 June 1956; the NNSC agreed to withdraw its teams, but recommended that it retain right to dispatch NNITs when desired.

15. Letter from Senior Member, UNCMAC to NNSC, dated 8 June 1956, stated: it expects NNITs to be withdrawn to DMZ by 9 June 1956.

16. At the 72nd MAC meeting on 7 June 1956, the MAC was unable to agree on the proposal of the NNSC stated in the 5 June 1956 letter from the NNSC.

17. Letter from Senior Member, UNCMAC, dated 9 June 1956 stated: the UNC regrets the MAC disagreement on the NNSC proposal of 5 June 1956, and stated the MAC records and the UNC letter of 8 June 1956, constitute UNC reply to NNSC letter of 5 June 1956.

조의 활동을 규정하는 부분의 조항을 국제연합군측 통제지역내에서 실시하는 것을 중지시키는 바이다.

13. 군사정전위원회 국제연합군측 수석위원이 중립국감독위원회에 발송하였으며 공동일직 장교를 통하여 전달된 1956년 5월 31일부 서한은 다음과 같이 지적하였다: 조항의 실시 중지는 약 1주일 내로 실시할 것이며 시찰 소조들이 그때까지 철수할 것을 요청한다.
14. 중립국감독위원회가 군사정전위원회에 보낸 1956년 6월 5일부의 서한을 통하여: 중립국감독위원회는 그의 시찰 소조들을 철수시키는 데 동의하였으나 요망될 때에는 시찰 소조를 파견할 수 있는 권리를 보류한다고 건의하였다.
15. 군사정전위원회 국제연합군측 수석위원이 중립국감독위원회에 보낸 1956년 6월 8일부의 서한은 다음과 같이 지적하였다: 국제연합군측은 중립국 시찰 소조들이 1956년 6월 9일까지 비무장지대로 철수할 것을 기대한다.
16. 1956년 6월 7일에 진행된 군사정전위원회 제72차 회의에서 군사정전위원회는 중립국감독위원회가 동 위원회의 1956년 6월 5일부 서한에서 언급한 바 제의에 대하여 합의를 볼 수 없었다.
17. 군사정전위원회 국제연합군측 수석위원이 보낸 1956년 6월 9일부 서한은 다음과 같이 지적하였다: 국제연합군측은 군사정전위원회가 1956년 6월 5일부 중립국감독위원회의 제의에 합의를 못 본 것을 유감으로 생각하며 군사정전위원회의 회의록과 국제연합군의 1956년 6월 8일부 서한은 중립국감독위원회의 1956년 6월 5일부 서한에 대한 국제연합군측의 답변이 된다.

18. Letter from the NNSC to the MAC, dated 9 June 1956 in which the NNSC acknowledged receipt of letters from both sides and stated, they informed NNITs to suspend activities and return to Panmunjom at 0001 hours on 9 June 1956.

28. Agreement between the United Nations Command and the Korean Peoples Army on Opening Some Area of the DMZ (West Side, 17 November 2000)

1. Both sides shall open some area of the MDL and the DMZ where the Seoul-Sinuiju railroad and the Munsan-Kaesong highway run through and put the area under the administration of the south and north in accordance with the Armistice Agreement.

2. Both sides shall make North and South militaries settle through negotiations the technical and business-like matters related to the opening of some area within the DMZ and the military matters arising in the area over which the south and north exercise administration in accordance with the Armistice Agreement.

3. This agreement shall become effective upon the date of approval at the Panmunjom General Officer Talks.

17 November 2000

MICHAEL M. DUNN
Major General
United Nations Command
Representative

PAK RIM-SU
Senior Colonel
Korean People's Army
Representative

18. 중립국감독위원회가 군사정전위원회에 보낸 1956년 6월 9일부의 서한은 쌍방으로부터 서한을 접수하였음을 확인하면서 동 감독위원회는 중립국 시찰 소조들에게 활동을 중지하고 1956년 6월 9일 0001시까지 판문점에 돌아올 것을 통고하였다고 지적하였다.

28. 비무장지대 일부구역 개방에 대한 국제연합군과 조선인민군간 합의서

1. 쌍방은 정전협정에 따라 서울－신의주간 철도와 문산－개성간 도로가 통과하는 군사분계선과 비무장지대의 일부구역을 개방하여 그 구역을 남과 북의 관리구역으로 한다.
2. 쌍방은 비무장지대 안의 일부구역 개방과 관련된 기술 및 실무적인 문제들과 남과 북의 관리구역에서 제기되는 군사적인 문제들을 정전협정에 따라 남과 북의 군대들 사이에 협의 처리하도록 한다.
3. 본 합의서는 판문점 장성급 회담에서 비준한 날로부터 효력을 발생한다.

2000년 11월 17일

마이클 엠. 던	박 림 수
소장	대좌
국제연합군측 대표	조선인민군측 대표

29. Agreement between the United Nations Command and the Korean Peoples Army on Opening Some Area of the DMZ (East Side, 12 September 2002)

1. Both sides shall open some area of the MDL and the DMZ where the Donghae railroad (Jeojin-Onjong-Ni) and the Songhyon-Ni-Gosong highway run through and put the area under the administration of the south and north in accordance with the Armistice Agreement.

2. Both sides shall make North and South militaries settle through negotiations the technical and business-like matters related to the opening of some area within the DMZ and the military matters arising in the area over which the south and north exercise administration in accordance with the Armistice Agreement.

3. This agreement shall become effective upon the date of approval at the Panmunjom General Officer Talks.

12 September 2002

JAMES N. SOLIGAN
Major General
United Nations Command
Representative

RI CHAN BOK
Colonel General
Korean People's Army
Representative

29. 비무장지대 일부구역 개방에 대한 국제연합군과 조선인민군간 합의서

1. 쌍방은 정전협정에 따라 온정리—저진 사이 철도와 고성—송현리 사이의 도로가 지나가는 군사분계선과 비무장지대의 일부구역을 개방하여 그 구역을 북과 남의 관리구역으로 한다.
2. 쌍방은 비무장지대안의 일부구역 개방과 관련한 기술 및 실무적인 문제들과 북남관리구역에서 제기되는 군사적인 문제들을 정전협정에 부합되게 북과 남 군대들 사이에 협의 처리하도록 한다.
3. 본 합의서는 판문점 장령급 회담에서 비준한 날부터 효력을 발생한다.

2002년 9월 12일

제임스 엔. 솔리간	리 찬 복
소장	상장
유엔군측 대표	조선인민군측 대표

5. Report by the Commander of the United Nations Unified Command in Korea pursuant to United Nations Security Council Resolution 84 (1950)

(For the period : 01 January 2016~15 May 2018)

5. 유엔 안전보장이사회 결의 제84호(1950)에 따른 주한 유엔 통합사령부 사령관 보고서
(보고기간 : 2016년 1월 1일~ 2018년 5월 15일)

• 유엔안전보장이사회에 보고하는 유엔군사령관 보고서의 일부를 소개하였습니다.

요약

1. ~ 5.

유엔군 사령관의 평가

I. 지정학적 배경

6. ~ 8.

II. UNC 중요 성과

9. ~ 11.

III. UNC 2018~2019, 향후 추진 방향

12. ~ 13.

법적 근거

14.

유엔군사령부 구성

15. ~ 17.

유엔군사령부 본부

I. 행정

18. ~ 19.

II. 인원 및 편성

20. ~ 21.

III. UNC Operations, Training & Conferences

22. Commercial fishing in the West (Yellow) Sea is a major economic endeavor for the ROK, DPRK, and the PRC. There are competing interests among commercial fishing enterprises for scarce resources in the area. The search for new fisheries led commercial fishing vessels from the PRC to enter the HRE. The Armistice Agreement opens the HRE to civil shipping by the sides who are signatories to the Agreement and directed the establishment of rules to govern activities in the HRE. An October 1953, Subsequent Agreement on Rules for Civil Shipping in the HRE and Related Matters levied specific requirements upon civil shipping activities.

23. In June 2016, vessels, from the PRC, committed sustained incursions into the HRE, a military-controlled area, to engage in unauthorized commercial fishing and other undetermined civil shipping activities. The UNCMAC Secretariat (UNCMAC-S) advised the KPA of these incursions and notified them of its intent to employ civil police patrol boats in the HRE to prevent further unauthorized activities. In the first employment of civil police boats in the HRE since the signing of the Armistice in June 2016, the UNC joined with civil police provided by the ROK military to patrol areas in the HRE. Unauthorized vessels were dissuaded from entering the HRE and remained outside of the area. Annually, there are two fishing seasons: April-June and September-December. The UNCMAC-S sustained support to civil police operations in the HRE during each of the fishing seasons in 2017 and 2018. The patrols prevented unauthorized activities and achieved the intended reduction of tensions in the area with no reported incidents in 2017 or through 15 May 2018.

III. 유엔사 작전 및 훈련, 회의

22. 서해(황해)에서의 상업적 어로 활동은 한국, 북한, 중국의 주요 경제 활동에 해당한다. 해당 지역의 희소한 자원을 놓고 상업적 어업회사간 이익이 상충하고 있다. 새로운 어장을 찾으려는 노력으로 인해 중국의 상업 어선들이 한강하구에 진입하기 시작하였다. 정전협정에 따라 정전협정 서명 당사국이자 한강하구 내 활동에 관한 규칙을 규정할 책임이 있는 국가의 한강하구내 민용 선박 항행은 허용된다. 1953년 10월에 비준된 한강하구에서의 민용 선박 항행에 관한 규칙 및 관계 사항에서 민용 선박의 항행에 대한 구체적인 요구 사항들을 규정하고 있다.

• 정전협정 체결 이후 추가 합의 사항목록 #1 : 한강하구에서의 민용선박 항행에 대한 규칙 및 관계 사항

23. 2016년 6월, 중국 선박들은 군사통제지역인 한강하구를 지속적으로 침범하여 미승인된 상업적 어로 행위 및 미상의 활동을 실시하였다. 유엔사 군정위 비서처는 북한군에 이러한 침범에 대해 알리고 추가적인 미승인 활동을 방지하기 위하여 한강하구내에 민사행정 경찰 선박들을 파견할 의도가 있음을 통보하였다. 2016년 6월, 정전협정 서명 이래 최초로 실시된 한강하구내 민정경찰 작전 당시 유엔사는 한국군이 제공한 민정경찰 인원들과 한강하구 순찰을 실시하였으며 미승인된 선박들이 한강하구에 들어오지 못하고 외부에만 머물도록 조치하였다. 성어기는 매년 4~6월, 9~12월로 연 2회 돌아오며 군정위 비서처는 2017년과 2018년 성어기에도 한강하구 민정경찰 작전을 지속 지원하였다. 민정경찰대는 해당 지역에서 미승인 활동을 방지하고 의도한 대로 긴장 완화를 달성하였으며, 2017년뿐 아니라 2018년 5월 15일 현재까지 단 한 건의 사건도 발생하지 않았다.

January 2017, the UNCMAC Secretariat verified the location of the loudspeaker systems. The UNC periodically inspected the speaker sites and monitored the content of broadcasts to ensure they were neither provocative nor offensive. Following the 27 April 2018 inter-Korean Summit, the ROK Military was directed to remove the loudspeaker systems in/near the DMZ. All systems were removed in early-May 2018.

27. The UNC Naval Component hosted the third annual Mine Countermeasures (MCM) Symposium from 11 to 14 October 2016 in Seoul. This was the first time the Symposium and the multinational mine exercise, CLEAR HORIZON, were combined as a larger Combined Mine Warfare (MIW) Event. The linkage resulted in a two-fold increase in attendance and a very robust discussion on capabilities that may be employed in the defense of the ROK. Over 40 MIW/MCM subject matter experts and 40 UNC Liaison Officers, Defense Attachés and Korea Planners from 11 nations participated. Presentations were conducted in USFK's Combined Battle Staff Room/Command Center Seoul. This is also the first time since the Korean War that members from Thailand and the Philippines returned to Korea to attend an exercise. The primary emphasis was for UNC Member States to send MCM assets during one of their MIW exercises and participate in the MIW during Exercises UFG and KR/FE.

28. The UNC Assistant Chief of Staff, Logistics (U4) hosted the fourth annual UNC Logistics Summit and first UNC Force Flow Table Top Exercise on 6-7 April 2016 in Seoul. This event was attended by participants from 13 nations and included 20 General Officers / Flag Officers from the ROK, U.S. and Member States. This conference focused on the UNC Commanding General's revitalization efforts as

24. 비슷한 노력의 일환으로 한국 정부는 파괴적 어획 행위를 예방하고 생태계를 재생하기 위해 서해(황해)에 인공어초를 설치하는 것과 관련하여 유엔사와 협의하였다. 어초 설치 장소가 적대 쌍방 군대가 서로 근접할 수 있는 곳이기 때문에 유엔군 사령관은 해당 계획을 검토하고 위험 완화 방안에 대한 지침을 제공하였으며 설치를 승인하되 유엔사 참관단이 이를 감시하도록 하였다. 해당 활동은 2016년 7월말에 무사히 완료되었다.

25. 2016년 4월 28일, 미 국방부 전쟁포로/실종자 확인국(DPAA)이 북한에서 발굴한 카투사 유해 15구를 한국으로 송환하는 행사가 한국 국방부장관과 유엔군 사령관 주관으로 실시되었다. 한국 국방부는 유해발굴감식단이 2015년 11월 한국에서 발견한 미국/유엔사 인원으로 추정되는 유해 2구를 송환하였다. 해당 미국/유엔사 유해는 유엔사를 거쳐 DPAA로 송환되었다. DPAA 국장은 용산 미군기지 나이트필드 연병장에서 실시된 송환식에도 참석하였다.

26. 2016년 11월에 실시된 여러 회의에서 한국 합참의장은 유엔군 사령관에게 한국 정부가 비무장지대(DMZ) 내외에 심리전 방송용 확성기를 설치하여 DMZ 너머로 방송을 실시하고자 한다는 계획을 전달하였다. 한국 합참은 확성기를 북한의 공세적 대남 심리전 방송에 대항하여 사용할 것이라고 설명하였다. 2016년 11월 15일, 유엔군 사령관은 고정형 심리전 확성기 24개, 이동식 심리전 확성기 16개의 설치를 승인하였다. 2016년 11월부터 2017년 1월까지 유엔사 군정위 비서

January 2017, the UNCMAC Secretariat verified the location of the loudspeaker systems. The UNC periodically inspected the speaker sites and monitored the content of broadcasts to ensure they were neither provocative nor offensive. Following the 27 April 2018 inter-Korean Summit, the ROK Military was directed to remove the loudspeaker systems in/near the DMZ. All systems were removed in early-May 2018.

27. The UNC Naval Component hosted the third annual Mine Countermeasures (MCM) Symposium from 11 to 14 October 2016 in Seoul. This was the first time the Symposium and the multinational mine exercise, CLEAR HORIZON, were combined as a larger Combined Mine Warfare (MIW) Event. The linkage resulted in a two-fold increase in attendance and a very robust discussion on capabilities that may be employed in the defense of the ROK. Over 40 MIW/MCM subject matter experts and 40 UNC Liaison Officers, Defense Attachés and Korea Planners from 11 nations participated. Presentations were conducted in USFK's Combined Battle Staff Room/Command Center Seoul. This is also the first time since the Korean War that members from Thailand and the Philippines returned to Korea to attend an exercise. The primary emphasis was for UNC Member States to send MCM assets during one of their MIW exercises and participate in the MIW during Exercises UFG and KR/FE.

28. The UNC Assistant Chief of Staff, Logistics (U4) hosted the fourth annual UNC Logistics Summit and first UNC Force Flow Table Top Exercise on 6-7 April 2016 in Seoul. This event was attended by participants from 13 nations and included 20 General Officers / Flag Officers from the ROK, U.S. and Member States. This conference focused on the UNC Commanding General's revitalization efforts as

처는 해당 확성기들의 위치를 검증하였다. 유엔사는 정기적으로 확성기 설치 장소를 점검하고 방송 내용이 도발적이거나 공세적이지 않은지 확인하였다. 2018년 4월 27일 실시된 남북정상회담 이후 한국군에 DMZ내/인근 확성기를 철거하라는 지시가 있었으며 2018년 5월초, 모든 확성기가 철거되었다.

27. 유엔사 해군구성군사령부는 2016년 10월 11~14일 서울에서 제3회 연례 기뢰대응책 심포지움을 개최하였다. 다국적 기뢰전 연습인 클리어 호라이즌과 심포지움을 통합하여 대대적인 연합 기뢰전 행사를 개최한 것은 처음이었으며, 두 행사를 연계하여 실시한 덕에 참석자 수가 두 배로 증가하였고, 한국 방어에 활용될 수 있는 역량들에 대한 활발한 토의가 이루어졌다. 기뢰전/기뢰대응책 전문가 40여 명과 유엔사 연락장교, 국방무관 및 11개국의 한국 담당 계획관 40여 명이 행사에 참여하였으며, 주한미군사 서울지휘소 CBS룸에서 발표가 실시되었다. 또한, 6.25한국전쟁 이후 처음으로 태국, 필리핀 인원들이 연습 참가를 위해 한국을 방문하였다. 기뢰전 연습시 유엔사 회원국들의 기뢰대응 자산 파견 및 UFG및 KR/FE연습간 기뢰전 훈련 참여 방안이 중점적으로 논의되었다.

28. 유엔사 군수참모부장(군참부장)은 2016년 4월 6~7일 서울에서 제4회 연례 군수회의 및 제1회 유엔사 전력전개 TTX를 주관하였다. 해당 행사에는 한 · 미 · 유엔사 회원국 장성 20명을 비롯하여 13개국 인원이 참석하였으며, 주요 토의 주제는 유엔군 사령관의 유엔사 재활성화 노력 및 다국적 군수 통합과 관련된 진행

well as developments related to multinational logistics integration. Discussions were conducted on potential requests for Member state logistics units and the processes for bringing those units into the coalition. These presentations enhanced the understanding of senior Member state officers on UNC force flow and reception, staging, and onward movement processes as well as the necessity of each UNC Member state to obtain the requisite bi-lateral legal agreements to support the ROK during contingency. During 2016, the U4 also hosted a conference on the integration of UNC national support elements into the theater logistics construct, and tested many aspects related to UNC logistics integration during Exercises UFG and KR/FE.

UNITED NATIONS COMMAND MILITARY ARMISTICE COMMISSION

I. Administration

29. Monthly, the Senior Member led the UNCMAC Advisory Group, in fulfillment of its Armistice responsibilities, and directed nine UNCMAC-led Special Investigations of suspected violations of the Armistice Agreement and reviewed the results of five incident investigations conducted by the ROK military that had Armistice implications. The UNC Commander hosted the first former UNCMAC Senior Members' Forum In December 2016. This conference presented a unique opportunity to draw upon the special experience of 18 ROK General Officers who served as the UNCMAC Senior Member to discuss their perspectives on improving Armistice maintenance. Follow-on conferences were held bi-annually in 2017 and the next forum is scheduled for June 2018.

상황 등이었다. 유엔사 회원국 군수 부대에 대한 지원 요청 가능성 및 해당 부대들의 동맹 통합 절차 등에 대한 토의도 실시되었다. 이러한 발표 및 토의를 통해 유엔사 전력 전개 및 RSOI 절차뿐 아니라 유엔사 회원국들이 각각 우발시 한국을 지원하는 데 필요한 한국과의 양자간 법적 합의를 맺는 것의 필요성에 대해서도 회원국 소속 고위급 장교들의 이해를 도모할 수 있었다. 유엔사 군참부장은 또한 2016년에 유엔사 전력 제공국들의 각 국별 지원 요소들을 전구 군수 체계에 통합하는 것과 관련한 컨퍼런스를 주관하여 UFG 및 KR/FE 연습시 유엔사의 군수 통합과 관련된 여러 분야를 점검하였다.

유엔군사령부 군사정전위원회

I. 행정

29. 수석대표는 매달 정전 책임 이행을 논의하는 유엔사 군정위 자문단 회의를 주관하였으며 정전협정 위반으로 의심되는 9건의 사건에 대해 유엔사 군정위 특별조사를 지시하고 정전 유지와 관련된 5건의 사건에 대한 한국군 자체 조사 결과를 검토하였다. 2016년 12월에는 유엔군 사령관 주관으로 제1회 역대 유엔사 군정위 수석대표 포럼이 개최되었다. 해당 회의는 수석대표로 근무했던 한국군 장성 18명이 자신들의 특수한 경험에 기반하여 정전 유지의 개선을 위한 의견을 공유하는 귀중한 자리였다. 첫 회의 이후 2017년에 두 차례 더 실시되었으며 차기 포럼은 2018년 6월로 계획되어 있다.

30. Armistice maintenance includes the ability to continue to communicate with the opposing side. The Military Armistice Commission Headquarters Area (MACHA) provides a location for dialogue and exchanging information between the UNC and KPA. Within the MACHA, the UNCMAC Secretariat operates a Joint Duty Office staffed 24 hours a day. The UNCMAC Joint Duty Office in Panmunjeom delivered 40 messages in 2016, 46 in 2017 and 30 through 15 May 2018 to the KPA on operations issues, exercise notifications, and administrative and maintenance matters. The significant increase in messages in the first four months of 2018 is attributed to support for the Third Inter-Korean Summit that was held on 27 April 2018. The KPA did not routinely respond to UNC messages. There were two exceptions: on 17 February 2017, the KPA responded to a request from the ROK government to repatriate five DPRK fishermen rescued at sea and on 13 November 2017, the KPA signaled to request the re-transmission of a message providing the status of a KPA soldier who defected to the ROK through Panmunjeom. Maintaining compliance with the Armistice Agreement also includes controlling access into the southern portion of the DMZ and authorizing crossings of the MDL in the two Transportation Corridors that traverse the DMZ; inspecting UNC GP/OP along the DMZ, HRE, and in the NWI; training military forces on their responsibilities to ensure Armistice compliance; conducting "right-to-flight" helicopter operations in the DMZ; observing artillery live-fire exercises on the NWI; and assisting or conducting Special Investigations of actual or alleged Armistice violations. To increase transparency of actions, the UNCMAC Secretariat includes accredited liaison officers from UNC Member States as team members and invites the Swedish and Swiss Neutral

30. 정전 유지를 위해서는 상대측과의 통신을 유지할 수 있는 능력이 구비되어야 한다. 군사정전위원회 본부구역(군정위 본부구역)은 유엔사와 북한군간 대화와 정보 교환을 위한 장소이다. 유엔사 군정위 비서처는 군정위 본부구역 내에 공동일직실을 24시간 운영하고 있다. 판문점내에 위치한 유엔사 군정위 공동일직실이 작전, 연습 통보, 행정 및 관리 등의 사안에 대해 전달한 통지문은 2016년 40건, 2017년 46건, 2018년은 5월 15일 기준 30건에 달한다. 2018년 초 4개월간 통지문이 대폭 증가한 것은 2018년 4월 28일 실시된 제3차 남북정상회담 지원에 기인한 것이었다. 북한군은 통상적으로 유엔사의 통지문에 대한 답변을 제공하지 않았다. 그러나 2017년 2월 17일에는 한국 정부가 해상에서 구조한 북한 어민 다섯 명의 송환을 요청한 것에 답변을 제공하였고, 2017년 11월 13일에는 판문점에서 한국으로 귀순한 북한군의 상태에 관한 통지문을 재전송해 줄 것을 요청하는 신호를 보내왔던 두 건의 예외적인 사례가 있었다. 그 밖에도 정전협정 준수를 위해 남측 비무장지대 출입 통제, 비무장지대를 관통하는 관리구역을 통한 군사분계선 통과 승인, 비무장지대, 한강하구, 서북도서의 유엔사 GP/OP 점검, 비무장지대내 '비행 권한' 행사를 위한 헬기 비행, 서북도서 해상사격훈련 참관, 정전협정 위반 및 위반 의심 사건에 대한 특별조사 등을 실시하고 있다. 이러한 활동의 투명성 강화를 위해 유엔사 군정위 비서처는 유엔사 회원국 연락장교들을 참가시키며, 모든 정전유지 활동간 공정한 참관을 위해 중립국감독위원회(중감위)의 스웨덴, 스위스 대표단의 참여를 요청한다.

NNSC Delegations to send officers as impartial observers during Armistice maintenance activities.

II. Operations

31. The Armistice Agreement solely authorizes UNCMAC to grant access to the southern portion of the DMZ and to cross the MDL-a robust, daily duty. In 2016, the UNCMAC Secretariat' s Transportation Corridor Control Teams processed transit requests and approvals for the round-trip movements of 14,896 passengers and 11,579 vehicles through the DMZ to the Kaesong Industrial Complex, an inter-Korean joint venture that is located north of the DMZ in North Korea, and employed approximately 53,000 North Koreans who work at over 120 South Korean-managed businesses. On 11 February 2016, the ROK Government closed the Kaesong Industrial Complex and ordered the departure of all ROK citizens. With less than 24 hours prior notice, the remaining ROK 250 citizens were accounted for and returned to South Korea. The Transportation Corridors sat dormant until early 2018. In a sign of inter-Korean cooperation with support from the international community, approximately (00) athletes, cultural performers, support staff and a high-level delegation led by DPRK Presidium President Kim, Yong-nam entered the ROK through the Transportation Corridors to participate in the 2018 PyeongChang Winter Olympic Games and the Paralympics.

32. The UNCMAC Secretariat conducted 42 inspections of 143 UNC guard and observation posts in 2016, 49 in 2017 and 41 in the first five month of 2018; led 11 monthly "right-to-flight" operations in each 2016 and 2017 and the first four months of 2018; in 2016, led four observations of artillery live fire exercises on the Northwest

II. 작전

31. 정전협정에 따라 유엔사군정위는 남측 DMZ 출입 및 군사분계선(MDL) 통행을 승인하는 권한을 가지며 이러한 승인은 일상적으로 활발하게 이루어지고 있다. 2016년 유엔사 군정위 남북관리구역 통제반은 인원 14,896명 및 차량 11,579대의 DMZ 통행 및 개성공단 출입을 승인하였다. 개성공단은 DMZ 이북 북한지역에 남북 합작으로 설립되었으며, 120여 개의 한국 기업이 입주하고 약 53,000명의 북한 근로자가 근무하였다. 2016년 2월 11일, 한국 정부는 개성공단 가동을 중단하고 모든 한국인의 철수를 지시하였다. 개성공단에 남아있던 한국인은 약 250명으로 파악되었으며 이들은 24시간 안에 한국으로 복귀하였다. 이후 2018년 초까지 남북관리구역은 사실상 휴면상태였다가 2018년 평창동계올림픽 및 패럴림픽을 앞두고 국제사회의 지지를 받아 남북 협력을 보여주는 조치로 김영남 북한 최고인민위원회 상임위원장이 이끄는 북한 선수단, 예술단, 관계자 및 고위급 대표단 (00)명이 남북관리구역을 통해 한국으로 넘어왔다.

32. 유엔사 군정위 비서처는 143개의 유엔사 GP/OP를 대상으로 2016년 42회, 2017년 49회, 2018년 첫 5개월간은 41회에 걸쳐 정기 점검을 실시하였다. H-128 권한 확인 비행은 2016년과 2017년, 2018년 첫 4개월에 11회 실시하였다. 백령도와 연평도에서 실시하는 서

Islands of Paengyong and Yonpyeong and the one exercise conducted in 2017; UNCMAC provided Armistice education on 62 occasions in 2017 and 111 presentations to all frontline ROK military units and to senior commanders and leaders at the Korea National Defense University and the Joint Forces Military University. Noteworthy is the fact that since the Armistice education program began in April 2014 until 4 May 2018, 300 presentations have been made. Educating front-line forces is a component of tension reduction.

III. Armistice Violations

33. The Korean Armistice is an agreement between the opposing military commanders that binds and governs their actions by the conditions and terms of the Armistice Agreement, whose articles and paragraphs shall remain in effect until expressly superseded by either mutually acceptable amendments or additions. To establish the DMZ, in July 1953 the Armistice Agreement directed that all military personnel, supplies, and equipment be withdrawn from the zone. In response to egregious KPA provocations and Armistice violations, UNC Commanders over time have made the deliberate decision to approve specific exceptions to the Armistice Agreement for the purpose of self-defense while preserving stability and preventing escalation of conflict.

34. During 2016, UNC observed activities on both sides of the DMZ that were actual or suspected violations of the provisions of the Armistice Agreement and the subsequent agreements to the Armistice Agreement. The vast majority of violations in the DMZ were minor in nature and included unauthorized access, unauthorized construction of fortifications, and the presence of

북도서 실사격훈련 참관은 2016년 4회, 2017년 1회 실시하였다. 정전 교육의 경우 전방 부대뿐 아니라 국방대, 합동대의 고위급 지휘관들을 대상으로 2017년 62회를 비롯해 총 111회 실시하였다. 정전 교육 프로그램이 2014년 4월에 시작된 이래로 2018년 5월 4일까지 300차례나 실시되었다는 점은 주목할 만하다. 전방 부대들을 대상으로 한 교육은 긴장 완화를 위한 조치의 일환이다.

III. 정전협정 위반

33. 한국 정전협정은 양측의 군 사령관들의 행동을 구속하고 관장하는 협정으로써, 쌍방의 군사령관들은 정전협정의 조항과 조건을 준수해야 한다. 정전협정은 상호 합의에 따른 개정이나 증보를 통해 명백히 대체되기 전까지 유효하다. 비무장지대를 수립하기 위해, 1953년 체결된 정전협정은 모든 무장 병력 및 물자와 장비가 비무장지대 내에서 철수할 것을 명시하였다. 북한군의 심각한 도발과 정전협정 위반에 대한 대응으로, 역대 유엔군 사령관들은 안정을 유지하고 분쟁고조를 방지하면서도 자위권을 행사할 수 있도록 정전협정에 대한 구체적인 예외사항을 승인해 왔다.

34. 2016년, 유엔사는 정전협정 및 정전협정 후속합의서 제반조항에 대한 실제 위반 사건이나 위반으로 의심되는 활동을 남측과 북측 비무장지대에서 공히 관측하였다. 비무장지대내에서 발생한 위반사항의 대다수는 무단출입, 비인가 진지 공사, 비인가 화기 반입 등의 경미한 사안들이었다.

unauthorized weapons.

35. Transparency of KPA activities in the northern half of the DMZ remained limited to the UNC. Until the KPA agrees to establish Joint Observer Teams prescribed by paragraphs 26 and 27 of the Armistice Agreement, complete and accurate accounting of KPA infractions will continue to be impeded.

36. In 2016, the UNCMAC Secretariat conducted nine Special Investigations. Of the nine investigations, five cited KPA violations, three cited ROK military violations, and one found no violation. Of the nine incidents, three involved KPA and ROK forces in proximity to each other. ROK violated the Armistice Agreement or Armistice Rules of Engagement in two of three incidents. This was considerably less than its 10 incidents in 2015; KPA Armistice Agreement violations were the same as in 2015. During 2017, the UNCMAC Secretariat led six Special Investigation Teams to investigate incidents. The ROK military investigated four incidents. The UNCMAC Secretariat maintains documentation on the results of the ROK investigations. Reflecting the apparent improvement in inter-Korean relations from the beginning of 2018, there have been no incidents requiring special investigations.

37. In July 2016, the UNC CDR expanded the UNC's investigative mechanisms to include designating the ROK Military to investigate potential Armistice violations that occur within their ranks and do not involve the KPA. This "cooperative approach" is intended to improve partnership and shared ownership with the ROK Military on Armistice affairs. ROK investigations report their findings to UNCMAC of (1) what occurred (the facts) and (2) what corrective action(s) are recommended. Should the UNCMAC Members accept

35. 유엔사가 북측 비무장지대에서 이루어지는 북한군 활동을 전적으로 투명하게 관측하는 것은 여전히 제한된다. 정전협정 제26항과 27항에 명시된 공동감시소조 활동에 북한군이 동의하지 않는 한 북한군의 위반사항을 완전하고 정확하게 파악하는 것은 앞으로도 어려울 것이다.
36. 2016년 유엔사 군정위 비서처는 9건의 특별조사를 실시했다. 이 가운데 5건은 북한군의 위반사항, 3건은 한국군의 위반사항, 나머지 1건은 위반사항이 아닌 것으로 판명되었다. 이 9건 중 3건은 북한군과 한국군이 서로 근접한 상태에서 사건이 발생하였다. 그리고 이 3건 중 2건에서 한국군은 정전협정과 정전교전규칙을 위반하였다. 이는 2015년 10건에 비해 크게 줄어든 것이며, 북한군의 정전협정 위반 건수는 2015년과 동일했다. 2017년, 유엔사 군정위 비서처는 6건의 특별조사를 실시하였으며, 한국군은 4건의 사건에 대해 조사를 실시하였다. 2018년 연초부터 이어지고 있는 남북관계 개선을 반영하듯 현재까지 특별조사를 요하는 사건은 발생하지 않았다.

37. 2016년 7월, 유엔군 사령관은 유엔사의 조사 메커니즘을 확대하여 북한군이 연루되지 않고 한국군만 관여된 정전협정 위반 추정 사건에 대한 조사를 한국군이 실시하도록 하였다. 이러한 '협력적 접근'은 정전협정 관련 사안 처리에 있어 한국군과의 파트너십을 강화하고 주도적인 역할을 한국군과 공유하기 위한 것이다. 한국군은 조사를 실시한 후, (1) 사건 개요(관련 사실), (2) 시정조치를 위한 건의안을 포함하는 조

the report, the UNCMAC Senior Member briefs the results to the UNC Commander. If UNCMAC recommends additional examination, the Senior Member directs further investigation (with UNCMAC-S or UNC legal assistance, if necessary), or directs a UNCMAC-S-led Special Investigation Teams (SIT) to investigate the incident and report.

38. Since the inception of the "cooperative approach," the UNCMAC Secretariat led 17 SITs to investigate incidents. The ROK military investigated nine incidents. The UNCMAC Secretariat maintains documentation on the results of the ROK investigations.

39. Significant KPA Armistice violations committed during 2016 included: operating an unmanned aerial vehicle in the DMZ on 13 January 2016; jamming Global Positioning System (GPS) signals during the period 31 March-6 April 2016; and, incursions into and conducting military training in the HRE on 4 and 6 October 2016. In 2017, the vast majority of Special Investigations conducted on North Korean-related incidents were defections to the ROK; both civilian and military; eight in total. The most significant incident involved a KPA soldier who defected to the ROK through Panmunjeom.

40. Armistice Agreement violations committed by the ROK military included two instances of firing warning shots at vessels from the DPRK on 8 February and 27 May 2016; the negligent discharge of a weapon within the DMZ on 3 April 2016; and the firing of warning shots while the DPRK conducted unmanned aerial vehicle within its territory on 10 April 2016. The significant event in 2017 warranting a special investigation was the ROK military kinetic response to unconfirmed aerial objects passing over the Demilitarized Zone.

사 결과를 유엔사 군정위에 보고한다. 유엔사 군정위 대표단이 이 보고서를 수락할 경우, 유엔사 군정위 수석대표는 그 결과를 유엔군 사령관에게 보고한다. 유엔사 군정위가 추가 조사를 건의할 경우, 수석대표는 추가 조사(필요시 유엔사 군정위 비서처나 유엔사 법무실의 지원 제공)를 지시하거나 유엔사 군정위 비서처가 주도하는 특별조사를 지시하여 보고토록 한다.

38. 이러한 "협력적 접근" 도입 후, 유엔사 군정위 비서처는 총 17건의 특별조사를 실시하였고, 한국군은 9건의 조사를 실시하였다. 유엔사 군정위 비서처는 한측 조사 결과보고서를 모두 보관하고 있다.

39. 북한군이 2016년 자행한 중대한 정전협정 위반사항은 다음과 같다. 2016년 1월 13일 비무장지대내 무인기(UAV) 운용, 2016년 3월 31일부터 4월 6일간 GPS 신호 교란, 2016년 10월 4-6일 한강하구내 무단 진입 및 군사훈련 실시. 2017년 실시된 북한군이 연루된 사건에 대한 특별조사의 대다수는 북한 귀순자 관련 사건들이었으며, 총 8건의 북한 민간인과 군인 귀순에 대한 조사가 이루어졌다. 이 중 가장 중대한 사건은 북한군 병사의 판문점 귀순이었다.

40. 한국군에 의한 정전협정 위반은 2016년 2월 8일과 5월 27일 북측에서 내려온 선박에 대해 한국군이 실시했던 두 차례의 경고사격과 2016년 4월 3일 비무장지대내에서 있었던 오발 사건, 2016년 4월 10일 북한이 남측 영토에서 운용한 UAV에 대한 경고사격 등이 있다. 2017년 특별조사를 요했던 중대 사건은 비무장지대를 통과하던 미상의 항체에 대해 한국군이 사격을 실시한 사건이었다.

IV. Armistice Activities In Support Of Military Exercises

41. The UNC routinely advises the KPA of major military exercises being conducted on the Korean Peninsula. The notifications are made in the spirit of transparency and to reduce the chance of a misunderstanding of intentions. On 7 March 2016, the UNCMAC Joint Duty Officer in Panmunjom notified the KPA of the exercise dates for exercises KR/FE 2016.[1)] The Spokesperson of the DPRK Foreign Ministry subsequently issued a statement denouncing the exercises. The UNC elected not to inform the KPA of KR/FE 2017 as the dates were public knowledge.

42. On 22 August 2016, the UNCMAC Joint Duty Officer in Panmunjom contacted the KPA to deliver a notification letter on exercise UFG 2016.[2)] The UNCMAC Joint Duty Officer and Duty Translator notified the KPA of the dates for the exercise. The UNC notified the KPA of exercise UFG 2017 on 18 August 2017.[3)] Two KPA soldiers witnessed and recorded (audio and video) the formal message being passed; the JDO and Duty Translator passed the UFG notification message verbally in English and Korean. No other KPA reaction to the notification.

43. On 3 February, 11 May, 26 July, 28 November 2016 and 27 February 2017, UNCMAC notified the KPA of planned ROK military artillery live-fire exercises on Paengyong and Yonpyeong Islands.[4, 5, 6, 7, 8)] The messages advised the KPA that the live-fire artillery exercises were routine training events. They also noted that the UNCMAC Secretariat would deploy international teams of observers to both islands to monitor the ROK training exercises to ensure compliance with recognized international training standards. Since the KPA

IV. 군사 연습 지원을 위한 정전협정 활동

41. 유엔사는 한반도에서 실시되는 주요 군사 연습에 대해 통상적으로 대북 통보를 실시한다. 이러한 통보는 투명성의 정신에 입각해서, 또한 의도에 대한 오해를 줄이기 위해 실시된다. 2016년 3월 7일 유엔사 군정위 공동일직장교는 판문점에서 2016년도 키리졸브/독수리연습(KR/FE)에 대한 대북 통보를 실시하였다.[1] 대북 통보 후 북한 외무성 대변인은 성명을 발표하고 이 연습을 비난하였다. 유엔사는 2017년 KR/FE에 대해서는 연습일자가 이미 일반에 공개되었기 때문에 북한에 따로 통보를 하지 않기로 결정하였다.

42. 2016년 8월 22일, 유엔사 군정위 공동일직장교는 판문점에서 2016년 UFG 연습 통보 서한을 전달하기 위해 북측에 연락을 취했다.[2] 이를 통해 유엔사 군정위 공동일직장교와 통역관은 연습일자를 북측에 통보하였다. 유엔사는 2017년 8월 18일 2017년 UFG 연습일자를 북측에 통보하였다.[3] 북한군 2명이 대북 통지문 전달 현장에 내려와 이를 지켜보면서 녹화를 실시하였다. 공동일직장교와 통역관은 UFG 연습 통보 대북 통지문을 영문과 국문으로 육성으로 전달하였다. 연습 통보 후 북한군측 반응은 없었다.

43. 2016년 2월 3일, 5월 11일, 7월 26일 및 11월 28일과 2017년 2월 27일, 유엔사 군정위는 한국군이 백령도와 연평도에서 실시할 포실사격 연습에 대해 북측에 통보하였다.[4, 5, 6, 7, 8] 이를 통해 이 포실사격 연습이 통상적인 훈련 활동임을 북측에 전달하였고, 유엔사 군정위 비서처에서 국제 참관단을 백령도와 연평도에 파견하여 한국군의 훈련이 국제 훈련 기준을 준수하는지 확인할 것임을 통보하였다. 북한군이 2013년 3

1) See Appendix 2 for Notification Letter, 7 March 2016.

1) 부록 2 대북통지문(2016. 3. 7.) 참조.

2) See Appendix 3 for Notification Letter, 22 August 2016.

2) 부록 3 대북통지문(2016. 8. 22.) 참조.

3) See Appendix 4 for Notification Letter, 18 August 2017.

3) 부록 4 대북통지문(2017. 8. 18.) 참조.

4) See Appendix 4 5 for Notification Message, 3 February 2016.

4) 부록 5 대북통지문(20176. 2. 3.) 참조.

5) See Appendix 5 6 for

stopped answering telephone calls from the UNCMAC Joint Duty Office in Panmunjom on 6 March 2013, the Joint Duty Officer and Duty Translator read the message in English and Korean verbally at the MDL. Usually, the KPA sent soldiers to witness and record (audio and video) the messages being passed. The live-fire exercises were conducted without incident on 4 February, 12 May, 28 July, 29 November 2016 and 28 February 2017.

UNITED NATIONS COMMAND - REAR

44. The UNC maintains a Rear Headquarters in Japan, with an Armistice mission to maintain the agreement regarding the Status of United Nations Forces in Japan (UN SOFA). In a contingency, UNC Rear will coordinate UNC Member State force flow through the seven designated UN Bases in Japan. The Commander of UNC Rear is an Australian senior officer, and the Deputy Commander and Force Representative is an officer from Canada. During 2016, the use of the UN SOFA to gain access to designated UN Bases was the highest rate in ten years. The Canadian HMCS Winnipeg, conducted a port visit to Yokosuka Naval Base, and the French Navy Ship Vendemaire visited White Beach naval base. A total of 15 aircraft from Australia, Canada, France, New Zealand, and Thailand conducted visits to Kadena and Yokota airbases. There were also eight overflight missions of Japan by aircraft from Australia and New Zealand. A total of 882 personnel were granted entry to Japan under the UN SOFA.

45. Additionally, UNC Rear hosts orientation tours of the UN designated bases in Japan for ROK Ministry of National Defense

월 6일 이후 판문점내 유엔사 군정위 공동일직실과 연결된 전화를 받지 않고 있기 때문에, 공동일직장교와 통역관이 군사분계선에서 영문과 국문 통지문을 직접 읽는 형태로 대북 통보를 실시하고 있다. 북한군은 통상 대북 통지문이 전달될 때 군인들을 보내 통지문 전달 현장을 확인하고 이를 녹화하고 있다. 한국군의 실사격 연습은 2016년 2월 4일, 5월 12일, 7월 28일, 11월 29일과 2017년 2월 28일 무사히 실시되었다.

유엔군사령부 · 후방

44. 유엔사는 일본에 후방사령부(유엔후방사)를 유지하고 있다. 유엔후방사의 정전협정 관련 임무는 일본내 유엔군의 지위에 관한 협정(유엔군 지위협정)을 유지하는 것이다. 우발상황 발생시 유엔후방사는 유엔사 회원국의 병력이 일본내 7개의 유엔사후방기지를 통과할 수 있도록 협조한다. 유엔후방사령관은 호주 고위급 장교가 맡고 있으며, 부사령관 겸 회원국 대표는 캐나다 장교가 맡고 있다. 2016년 유엔사후방기지 출입을 위해 유엔군 지위협정이 최근 10년 들어 가장 활발하게 활용되었다. 캐나다 위니펙 함이 요코스카 해군기지에 정박하였고, 프랑스 해군 방데미에르 함이 화이트비치 해군기지에 정박하였다. 호주, 캐나다, 프랑스, 뉴질랜드, 태국에서 총 15대의 항공기가 가데나와 요코타 공군기지를 방문하였다. 또한 호주와 뉴질랜드 소속 항공기가 총 8차례 일본 상공을 경유하는 비행을 실시하였다. 총 882명이 유엔군 지위협정에 의거하여 일본 방문 허가를 득하였다.

45. 또한 유엔후방사는 한국 국방부 장성들과 국회의원들을 위한 일본내 유엔사후방기지 견학방문을 주관한

Notification Message, 11 May 2016.

5) 부록 6 대북통지문(2016. 5. 11.) 참조.

6) See Appendix 6 7 for Notification Message, 26 July 2016.

6) 부록 7 대북통지문(2016. 7. 26.) 참조.

7) See Appendix 7 8 for Notification Message, 28 November 2016.

7) 부록 8 대북통지문(2016. 11. 28.) 참조.

8) See Appendix 9 for Notification Message, 27 February 2017.

8) 부록 9 대북통지문(2017. 2. 27.) 참조.

general officers and members of the National Assembly. These visits are designed to create a high level of awareness and understanding of the role of Japan and US Forces Japan in a contingency on the Korean Peninsula. In 2016, four orientation tours were conducted for the ROK military, with a total of 22 Korean senior officers hosted. Two tours for members of the ROK National Assembly were also conducted to create a high level of awareness and understanding of the supporting role of Japan, U.S. Forces Japan, and U.S. Pacific Command Forces in a contingency on the Korean Peninsula.

NEUTRAL NATIONS SUPERVISORY COMMITTEE

46. The Armistice Agreement established the NNSC to carry out the functions of supervision, observation, inspection, and investigation at places outside of the DMZ.[9]
47. The Swedish and Swiss NNSC Delegations resident in Panmunjeom, at the request of the UNC Commander, and as authorized by the NNSC Reaffirmation of Armistice Obligations and Expanded Contributions of 2016, observed UNCMAC activities that included the observation of 46 inspections of UNC GPs/OPs; participation in 10 "right-to-flight" operations; observation of three ROK military artillery live-fire exercises conducted on the Northwest Islands; and, observed nine UNCMAC-led special investigations of Armistice violations. Based on the observations, the NNSC Delegations provided an independent report on each such activity.
48. The NNSC also assisted UNCMAC in Armistice education and orientation programs, participating in a total of 50 such events directed at ROK frontline military units, students at the ROK Joint Forces Military University and the ROK National Defense

다. 이 유엔사후방기지 방문은 한반도 우발상황시 일본과 주일미군의 역할에 대한 인식 및 이해를 심도 있게 제고하기 위해 실시되고 있다. 2016년 한국군을 위한 유엔사후방기지 견학이 4차례 실시되었으며, 총 22명의 한측 고위급 장교들이 참여하였다. 한국 국회의원들을 위한 2차례의 견학이 별도로 실시되어 한반도에서 우발상황 발생시 일본, 주일미군, 미태평양사령부의 지원 역할에 대한 높은 수준의 인식과 이해가 제고되었다.

중립국감독위원회(중감위)

46. 정전협정은 비무장지대 외부에서 감독, 참관, 점검 및 조사 활동을 실시하도록 중감위를 설치하였다.[9)]

47. 스웨덴과 스위스의 중감위 대표단은 유엔군 사령관의 요청에 따라 판문점에 상주하고 있으며, 2016 중감위 정전 의무 및 확대 기여 재확인 각서를 통해 위임받은 바와 같이 유엔사 군정위 활동을 참관하고 있다. 총 46차례의 유엔사 GP/OP 점검과 10차례의 권한 확인 비행, 서북도서에서 실시된 3차례의 한국군 포실사격 연습, 정전협정 위반사항에 대한 9차례의 유엔사 군정위 특별조사를 참관하였다. 참관 후 중감위 대표단은 각각의 활동에 대해 독자적인 보고서를 작성하였다.

48. 중감위는 유엔사 군정위의 정전협정 교육 프로그램도 지원하고 있으며, 한국군 전방부대와 합동군사대학교 및 국방대학교에서 실시된 교육에 50차례 참여하였다. 또한 판문점에 위치한 중감위 본부를 방문한

9) See Appendix 1 for a more detailed discussion of the Neutral Nations Supervisory Commission.

9) 중감위와의 토의 관련 세부사항은 부록 1 참조.

University. Furthermore, the NNSC Delegations, at their headquarters in Panmunjeom, received and gave presentation on the Armistice Agreement and the role of NNSC to some 78 visiting delegations, with a total of 1250 visitors.

49. The NNSC Delegations are tasked to observe specific military exercises and to provide an independent, impartial and objective assessment on these exercises. During 2016, the Swedish and Swiss NNSC Delegations observed two, major UNC forces' exercises and activities and one ROK military exercise. NNSC Delegations provided impartial and objective assessments in independent reports to CDR UNC and to the UNCMAC Senior Member of all observations. Based on their exercise observations, the NNSC concluded that the exercises were compliant with the Armistice Agreement and deterrent and defensive in nature.[10, 11]

총 78개 단체 1,250명에 대해 정전협정과 중감위의 역할에 대한 브리핑을 실시하였다.

49. 중감위 대표단은 특정 군사 연습을 참관한 후 독자적, 중립적, 객관적인 평가를 내리는 임무도 맡고 있다. 2016년 중감위 스웨덴/스위스 대표단은 유엔사측의 대규모 연습 2개와 한국군 연습 1개를 참관하였다. 중감위 대표단은 참관 결과에 대한 중립적이고 객관적인 평가보고서를 유엔군 사령관에게 제출하였고, 기타 모든 참관활동에 대한 결과를 유엔사 군정위 수석대표에게 제공하였다. 이러한 참관 활동을 통해, 중감위는 이러한 연습들이 정전협정을 준수하였으며 억제적이며 방어적인 성격의 연습인 것으로 결론내렸다.[10, 11]

10) See Appendix 8 for NNSC Report on Observations of KEY RESOLVE/FOAL EAGLE 2016.

10) 중감위의 2016 KR/FE 연습 참관 보고서는 부록 8 참조.

11) See Appendix 9 for NNSC Report on Observations of ULCHI FREEDOM GUARDIAN 2016.

11) 중감위의 2016 UFG 연습 참관 보고서는 부록 9 참조.

Ⅱ

평화
Peace

Chapter 2

1. On Peace

North Korea's continued provocations including nuclear tests and missile launches drove the Korean Peninsula to the brink of war in 2017. However, the thaw that began in 2017 has brought hope for peace and prosperity on the Korean Peninsula with the inter-Korean summits and the U.S.-North Korea summit in 2018.

At the same time, the political talks that were left unfinished among the relevant countries to find a peaceful resolution to the Korean issue within three months after the Korean Armistice Agreement are currently underway. The issues of an end-of-war declaration and a peace treaty on the Korean Peninsula are making progress.

Peace is literally a state of no war where everyone understands each other and are friendly and harmonious with one another without conflict or dispute.

President Moon Jae-in and Chairman Kim Jung-un planted a tree in April 2018 at Panmunjom for the peace and prosperity of the Korean Peninsula.

Although there have been some tensions and illegal provocations by North Korea since the signing of the Armistice Agreement, the Korean Peninsula has been in a state of stability thanks to UNCMAC and the NNSC, which have ensured the implementation of the Armistice

Chapter 2

1. 평화에 대하여

북한의 계속된 핵실험과 미사일 발사로 이어진 도발은 한반도를 전쟁의 위기로까지 몰고 갔습니다. 그러나 2017년에 시작된 화해의 분위기는 2018년 남북정상회담과 북미정상회담으로 이어졌고 한반도에 평화와 번영에 대한 희망을 가져다주었습니다.

6.25전쟁 이후 정전협정 효력이 발휘하고 3개월 이내 관련국가간의 정치회담으로 한반도 문제를 평화적으로 해결하려다가 아직까지 마무리하지 못한 정치적 회담이 진행되고 있습니다. 한반도에 대한 종전선언, 평화협정 등이 논의되고 있습니다.

평화는 사전적으로 전쟁을 하지 않는, 즉 분쟁과 다툼이 없이 서로 이해하고 우호적이며 조화를 이루는 상태입니다.

문재인 대통령과 김정은 국무위원장은 2018년 4월 판문점에서 한반도의 평화와 번영을 위하여 한 그루의 나무를 심었습니다.

평화와 번영의 나무
The Tree of Peace and Prosperity

정전협정체결 이후 한반도에서는 다소의 긴장과 북한의 불법적인 도발이 있었지만 그럼에도 불구하고 한반도에서의 정전상태와 안정을 유지할 수 있었던 것은 군사정전위원회와 중립국감독위원회의 정전협정이행과

Agreement, and our national strength together with the ROK-U.S. alliance and the UNC with its Sending States. It is this strong security that has enabled South Korea to work vigorously and achieve peace and prosperity. Since the Armistice Agreement in 1953, we have held numerous discussions with the North that have led to the 1991 Inter-Korean Basic Agreement, the June 15 Summit Declaration in 2000, the 10 April Joint Declaration in 2007, and the Summit Declarations and the Comprehensive Military Agreement in 2018, and our relentless commitment for peace and prosperity continues today.

At the inter-Korean summit on April 27, 2018, the two sides agreed to reduce military tensions and build confidence by transforming the DMZ into a peace zone and the West Sea as a peace zone and regularizing military talks between the two Koreas. To accomplish this, we held several military and civilian talks between the two Koreas, and as a result, we were able to draw up a military consensus for the implementation of Panmunjom Declaration on September 19th.

The core spirit of the CMA is to thoroughly implement the Panmunjom Declaration in the military domain based on the common understanding that easing military tension and building confidence on the Korean Peninsula is integral to securing lasting and stable peace(CMA Overview by ROK MND).

First, South and North Korea agreed to completely cease all hostile acts against each other in every domain, including land, air and sea. The two sides have stopped all hostilities against each other since November 1, 2018. On ground, the two sides agreed to cease all live-fire artillery drills and field training exercises at the regiment level and above within 5km from the MDL. At sea, the two sides agreed to cease all live-fire and maritime maneuver exercises within the zone north of Deokjeok-do and south of Cho-do in the West Sea, and within the zone

이를 힘으로 뒷받침한 한미동맹, 그리고 유엔전력제공국을 포함한 우리의 국력이었습니다. 이러한 안보를 바탕으로 대한민국은 끊임없는 노력을 통해 평화와 번영을 성공리에 이룩하여 왔습니다. 1953년 정전협정 이후 1972년 7.4남북공동성명, 1991년 남북기본합의서, 2000년 6.15정상선언, 2007년 10.4공동선언, 그리고 2018년 남북정상회담 및 군사분야 기본합의서 등 수없이 많은 협의를 해 왔으며, 지금도 평화와 번영을 위하여 노력하고 있습니다.

2018년 4월 27일 남북정상회담시 양측은 DMZ를 평화지대로 하고 서해해상을 평화수역으로 하며 남북간 군사회담을 정례화 함으로써 군사적 긴장을 완화하고 신뢰를 구축하고자 합의하였습니다. 이를 이행하기 위하여 수 차례 남북 장성급 및 실무급 군사회담을 하였고, 그 결과 2018년 9월 19일 판문점선언 이행을 위한 군사분야 합의를 이끌어 낼 수 있었습니다.

군사분야 합의의 주요내용은 한반도에서 군사적 긴장완화와 신뢰구축이 항구적이며 공고한 평화 보장에 필수적이라는 공통인식으로부터 판문점 선언을 군사적으로 철저히 이행하는 것입니다.*

*9.19 군사합의 관련 설명자료, 국방부

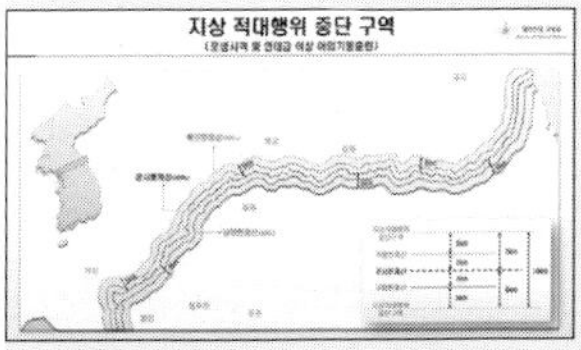

No Hostility Zone on Ground (No field training exercises at the regiment level and above)

첫째, 남과 북한은 지상, 해상, 공중을 비롯한 모든 공간에서 상대방에 대한 일체의 적대행위를 전면 중지하는 것입니다. 2018년 11월 1일부터 상호 적대행위를 중지하고 있습니다. 지상은 군사분계선 기준 남과 북쪽으로 각각 5km를 완충지대로 하여 포병사격 훈련과 연대급 이상 야외기동훈련을 중지하고 있습니다. 해상은 서해 남측 덕적도로부터 북측 초도, 동해 남측 속초로부터 북측 통천까지를 완충수역으로 설정하고 완충수역에서

north of Sokcho and south of Tongcheon in the East Sea. The two sides also agreed to install covers on the barrels of coastal artilleries and ship guns and close all gunports within the zones. In the air, the two sides agreed to designate No Fly Zones for all aircraft types including fixed-wing, rotary wing and UAVs above the MDL.

Second, South and North Korea agreed to devise substantive military measures to transform the DMZ into a peace zone. The two sides completely withdrew ten guard posts by December 31, 2018 that lie within 1km of each other as a preliminary measure to withdrawing all guard posts within the DMZ. The two sides agreed to demilitarize the Joint Security Area so as to turn it into a place of peace and harmony. Also, South and North Korea agreed to jointly recover remains in the vicinity of Arrowhead Hill in Cheorwon in the DMZ, where fierce battles were fought during the war.

Third, South and North Korea agreed to take military measures to prevent accidental military clashes and ensure safe fishing activities by turning the area around the Northern Limit Line in the West Sea into a maritime peace zone.

Fourth, South and North Korea agreed to devise military assurance measures necessary for invigorating exchanges, cooperation, contacts and visits. The two sides agreed to establish military measures to ensure passage, communication and customs in the East and West Transportation Corridors and to connect and modernize eastern/western railways and roads. The two sides agreed to discuss the issues related to permitting the use of Haeju Passage and Jeju Strait at the Inter-Korean Joint Military Committee. Also, South and North Korea agreed to devise military assurance measures for the shared use of Han River (Imjin River) Estuary.

Fifth, South and North Korea agreed to devise various measures for

는 포병 및 함포사격과 해상기동훈련을 중지하고 해안포 및 함포의 포구에 포신 덮개를 설치하고 포문은 폐쇄 조치를 하고 있습니다. 공중은 고정익, 회전익, 무인기, 기구 즉 기종별 군사분계선 상공에 비행금지구역을 설정하였습니다.

둘째, 비무장지대를 평화지대화로 변환을 위한 군사적인 대책을 마련하는 것입니다. 비무장지대내 모든 감시초소 철수를 위해 우선 시범적으로 1km 이내 근접한 10개의 감시초소를 2018년 12월 31일까지 철수를 완료하였습니다. 공동경비구역을 비무장화하여 평화와 화합의 장소로 전환하는 것입니다. 그리고 6.25전쟁시 격전지였던 비무장지대내 철원 화살머리고지 일대에서 남북이 공동으로 유해를 발굴하는 것입니다.

셋째, 서해 북방한계선 일대를 평화수역으로 만들고 시범적 공동어로구역을 설정하여 우발적인 군사적 충돌을 방지하고 안전한 어로활동을 보장하기 위한 군사적 대책을 강구하는 것입니다.

넷째, 남북간의 교류협력 및 접촉과 왕래를 활성화하기 위하여 군사적으로 여건을 보장하는 것입니다. 양측은 남북 관리구역에서 통행 · 통신 · 통관을 보장하고 동 · 서해선 철도 및 도로 연결과 현대화를 위한 군사적 보장대책을 강구하며 북측 선박들의 해주직항로 이용과 제주해협 통과 문제 등을 남북군사공동위원회에서 협의, 그리고 한강하구 공동이용을 위한 군사적 보장대책을 강구토록 하였습니다.

한강하구/ Han River Estuary

다섯째, 남과 북은 상호 군사적 신뢰구축을 위한 남북

mutual military confidence building including forming the Inter-Korean Joint Military Committee.

I recently visited a guard post removal site in the DMZ with the members of UNCMAC and its Secretariat. As I was looking at a site where a guard post once stood, my heart was filled up with emotions. There were tears of pride running down my heart as I was standing at the very place where my comrades before me had defended with their lives through ambush operations in long winter nights and reconnaissance and patrol operations in the mine-filled DMZ for decades.

On the other hand, I felt a sense of new hope sprung up as it occurred to me that this may be the rite of passage for Korea to become a more peaceful and prosperous country. A foreign officer who works at the Military Armistice Commission's Secretariat, said, as a soldier, he would remember it for a long time that he was at the forefront of the Korean Peninsula at this historic moment. The officer said that he was at the Berlin Wall when it fell, unifying West and East Germany, and that he could not hide his excitement that there would be a new hope on the Korean Peninsula.

Switzerland, despite being a beautiful country with plenty of water and mountains, suffered from poverty for a long time due to lack arable land and resources, and many of its young people became mercenaries all over Europe to find employment. Since 1474 when they first became mercenaries in France, as many as two million Swiss people served as mercenaries as foreign troops for centuries. Sometimes they had to fight each other for different countries that were at war with each other, and until 1848 when the Swiss

군사공동위원회 구성 등 다양한 조치를 강구하는 것입니다.

얼마 전 DMZ(비무장지대) 내 GP(감시초소) 철수 현장을 군사정전위원회 위원 및 비서처 인원과 함께 방문하였습니다. GP를 무너뜨린 현장을 보면서 만감이 교차함을 느꼈습니다. 이 비무장지대내에 있는 감시초소들은 오랜 기간 수많은 선배 전우 여러분들이 혹한의 추위 속에서 동지섣달 기나긴 밤에 매복 작전으로 지켜왔고, 수많은 지뢰가 흩어져 있는 비무장지대에 목숨을 걸고 수색정찰하면서 지켜왔던 바로 그 지역이기 때문에 가슴 속에는 뜨거운 감동이 벅차올랐습니다.

또 다른 희망은 평화와 번영을 위해 그리고 보다 나은 대한민국의 평화와 번영을 위해 이 또한 감내해야 하는 것 아닌가 하는 생각에 새로운 희망이 샘솟는 것을 느낄 수가 있었습니다. 그때 함께했던 군사정전위원회 비서처에 근무하는 한 외국군 장교는 이러한 역사적인 순간에 한반도의 최전선에 있음이 군인으로서 오래도록 기억될 것이라고 가슴 벅찬 감격을 전했습니다. 그 장교는 독일이 통일을 이루던 베를린 장벽이 무너진 그 장소에 그때 당시 함께했다면서 반드시 한반도에 새로운 희망이 있을 것이라며 감격을 감추지 못했습니다.

아름다운 나라 스위스는 산과 물은 풍족하나 부족한 경작지와 빈약한 자원으로 오랜 기간 가난에 시달렸고, 많은 젊은이 들은 직업을 찾아 유럽 등 주변 각국의 용병에 지원하였습니다. 1474년경 프랑스 용병으로 시작하여 수세기 동안 약 200만 명이 외국 군대에 용병으로 활동했습니다. 심지어 스위스 용병끼리 전쟁을 치르는 경우도 있었으며 1848년 정부가 법적으로 용병 취업을 금지할 때까지 약 50만 명이 여러 전투에서 목숨을 잃었습

government banned mercenary, some 500,000 Swiss people had lost their lives in numerous battles.

At a rocky cliff at the heart of Lucerne, one of the most beautiful tourist destinations in Switzerland, there is the statue of the Dying Lion of Lucerne, which portrays a lion impaled by a spear, protecting a shield bearing the emblem of the French monarchy of the Bourdon with his head and feet even at its dying breath.

When the French Revolutionary Army and French people stormed the Tuileries Palace in Paris on August 10, 1792, the royal guards ran away, leaving Louis XVI and the royal family behind in the palace. The Revolutionary Army demanded the Swiss Guards that were guarding the royal family to surrender, but the Swiss mercenaries shouted, "If we flee to live, who will entrust our brothers and children as their mercenaries in the future? " They decided to stay with the royal family and put up a fierce fight, but in the end, all 786 of them were killed by the Revolutionary Army. From this incident, however, the reputation of the Swiss mercenaries was built that they always stay loyal to their employers, along with the worldwide reputation of the Swiss people that they can be trusted and will get the job done. Earning trust takes sacrifices and dedication, just like those Swiss mercenaries who honored their word with their lives for their descendants.

Poland, which once served as an NNSC delegation nominated by North Korea, is currently participating in the NNSC activities regularly through the UNC side. It is located in the east-central part of Europe and boasts vast plains and abundant resources. In 1385, Queen Jadwiga, the monarch of the Kingdom of Poland, married Grand Duke Jogaila of Lithuania, creating the Polish-Lithuanian Commonwealth. By the 16th century, the Polish-Lithuanian Commonwealth became a

니다.

스위스에서 가장 아름다운 관광도시 중 하나인 루체른 시내 한복판 바위절벽에 있는 빈사의 사자상은 창에 찔린 사자가 자신의 모든 삶이 다하는 마지막 순간까지 프랑스 부르봉 왕가 문장이 새겨진 방패를 머리와 발로 지키고자 하는 형상을 표현하고 있습니다.

스위스 로체른 빈사의 사자상
The Lion of Lucerne

프랑스 혁명군과 시민들이 1792년 8월 10일 파리 튈르리 궁으로 몰려왔을 때 왕궁 수비대는 도망가고 루이 16세와 왕족들만 남았습니다. 혁명군은 왕실을 경호하고 있던 스위스 용병들에게 항복을 요구하였으나, 스위스 용병들은 "만약 우리가 살기 위해 도망간다면 누가 후대의 우리 형제 자식들에게 믿고 용병을 맡길 것인가" 라고 외치며 신의를 지키기 위하여 왕족들과 최후를 같이하고자 했습니다. 결국 용병들은 혁명군과 처절한 사투를 벌이다 786명이 전멸하였습니다. 이 사건을 통해 스위스 용병은 고용주에 대한 신의를 절대 저버리지 않는다는 전통을 만들었고, 또한 스위스인들은 모든 일을 믿고 맡길 수 있다는 범세계적인 신뢰를 얻었습니다.* 후손들을 위해 목숨으로 신의를 지킨 스위스의 용병과 같이, 신뢰를 얻기 위해서는 희생과 헌신이 따라야 한다고 생각합니다.

*뉴데일리. 2014. 스위스 용병 786명의 집단죽음 http://www.newdaily.co.kr/site/data/html/2014/08/22/2014082200028.html

한때 북한측 중립국감독위원회로 활동했고 현재 유엔군측 중립국감독위원회에 주기적으로 참가하고 있는 폴란드는 유럽의 중동부에 위치하고 있으며 넓은 평야와 자원이 풍부한 나라입니다. 1385년 왕위 계승자인 야드비가 여왕이 리투아니아의 야기에워 대공과 결혼하고 일종의 국가연합인 폴란드, 리투아니아 연합왕국을 형성하면서 16세기에는 유럽 최고의 곡창지대로서 군사력

military and economic powerhouse with the largest arable land in Europe.

In 1572, with no one to succeed the throne, Poland became a republic in which aristocrats elected the king, limiting the power of the monarchy. With the power concentrated on the aristocrats, Poland began to wane. At the end of the 18th century, Austria annexed some part of Poland, but Poland showed no physical resistance, which led to a discussion among its neighbors, Russia, Prussia, and Austria on how to divide and occupy the country. In the end, Poland was occupied by Russia, Prussia and Austria and ceased to exist as a sovereign nation for 123 years from 1772 to 1918. During the First World War, Poland gained independence thanks to President Wilson' s principle of self-determination, but again fell under Russia and Germany during the Second World War. From August to October 1944, the Warsaw Uprising against Germany cost hundreds of thousands of lives, and Poland became independent in the ruins after the World War II.

The rise and fall of a state shows the importance of choices by the people and the leaders. No matter how rich or powerful a country might be, it will cease to exist if it fails to defend and maintain its strength. Without a country, offering peace and happiness to the people will be challenging, if not impossible.

The security and peace of the Republic of Korea, despite the challenges in implementing the Armistice Agreement, has remained strong for a long period of time thanks to our unwavering faith in liberal democracy and national power that has backed it up. I believe our national power comes from not just the strength of the Republic of Korea, but also that of the U.S., our staunchest ally since 1953, and UNC with its Sending States that are committed to fighting shoulder to

과 경제력을 자랑하였습니다.

1572년 왕위 계승자 부재로 귀족들이 국왕을 선출하는 공화정이 등장했으며, 국왕의 권력이 귀족들에게 제한되면서 국력은 쇠퇴해졌습니다. 18세기 말 오스트리아가 일부 폴란드 영토를 병합하였음에도 어떠한 물리적 대응이 없자 국경을 접하고 있었던 러시아, 프러시아, 오스트리아는 폴란드 분할점령을 논의하기 시작하였고, 1772년부터 러시아, 프러시아, 오스트리아에 의한 국토 분할로 1918년까지 123년간 국가가 소멸되었습니다. 제1차 세계대전간 미국 윌슨 대통령의 민족자결주의 원칙에 따라 독립을 하였으나, 제2차 세계대전시 다시 러시아와 독일의 지배를 받았습니다. 1944년 8월에서 10월까지 독일에 항전하기 위한 바르샤바 봉기는 수십 만여 명의 희생이 뒤따랐으며 2차 세계대전 종료 후 폴란드는 폐허 속에서 독립을 하게 되었습니다.*

국가의 흥망성쇠는 당시의 국민과 지도자의 선택이 얼마나 중요한 것임을 잘 보여주고 있습니다. 부유하고 강력한 군사력을 보유하고 있더라도 이를 지키고 유지하지 못하면 국가를 유지할 수가 없고 국가가 존재하지 않는 곳에서 국민들의 평화와 행복은 찾기가 쉽지 않을 것입니다.

정전협정 이행의 어려움 속에서도 긴 시간 동안 대한민국이 평화와 안보를 지킬 수 있었던 것은 자유민주주의에 대한 확고한 신뢰와 이를 뒷받침해 온 국력입니다. 국력은 우리의 능력뿐만 아니라 지난 1953년 이후 함께 한 한미동맹과 한반도에 전쟁이 다시 일어나면 언제든지 와서 함께 싸우고자 다짐하고 있는 전력제공국 즉, 유엔군의 힘이었다고 생각이 됩니다. 그리고 무엇보다 군

*외교부. 2013. 폴란드 개황 http://www.mofa.go.kr/www/brd/m_4099/down.do?brd_id=15388&seq=358114&data_tp=A&file_seq=1

shoulder with us again at any time should there be another conflict on the Korean Peninsula. Above all, the people of the Republic of Korea who fulfilled their duties to defend the country either by being service members themselves or by supporting service members in their family, are the most important source of our national power. In short, we owe it to the dedication of the service members who came before us and the Korean people and the sacrifices of the ROK-U.S. alliance and the UN forces for the balance of power for peace on the Korean Peninsula.

To achieve a true peace and prosperity on the Korean Peninsula, we need to have commitment and strength as well as trust of the international community. There should never be another war on the Korean Peninsula and we, with unwavering faith and strength, must be willing to make sacrifices and dedication in our journey for peace and prosperity no matter how painful or slow it may be. Even at this very moment, we should not cease our efforts to maintain the balance of power to continue diplomatic dialogue to defend peace.

Senior Member, United Nations Command Military Armistice Commission
MG Shin Sang-bum, ROK Army

인과 군인의 가족 또는 친지가 함께 다양하게 국방의 의무를 수행해 온 대한민국 국민이 바로 주인공입니다. 선배 전우와 대한민국 국민들의 헌신 그리고 한미동맹과 유엔군의 희생을 바탕으로 한반도의 평화를 위한 힘의 균형을 이룩할 수 있었습니다.

한반도의 진정한 평화와 번영을 이룩하기 위해서는 이를 지키려는 우리의 의지와 힘 그리고 국제적인 신뢰가 필요합니다. 앞으로 한반도에는 더 이상 전쟁이 없어야 하며 평화와 번영을 이룩하고자 하는 과정이 고통스럽고 더디더라도 확고한 의지와 국력을 갖고 우리 모두가 희생과 헌신으로 나아가야 합니다. 오늘도 우리는 평화를 지키기 위한 외교적 대화를 지속할 수 있도록 힘의 균형을 유지해야 합니다.

유엔군사령부 군사정전위원회 수석대표
대한민국 육군소장 **신 상 범**

2. Pyongyang Joint Declaration

Moon Jae-in, President of the Republic of Korea and Kim Jong Un, Chairman of the State Affairs Commission of the Democratic People's Republic of Korea held the Inter-Korean Summit Meeting in Pyongyang from September 18 to 20, 2018.

The two leaders assessed the excellent progress made since the adoption of the historic Panmunjom Declaration, such as the close dialogue and communication between the authorities of the two sides, civilian exchanges and cooperation in many areas, and groundbreaking measures to defuse military tension.

The two leaders reaffirmed the principle of independence and self-determination of the Korean nation, and agreed to consistently and continuously develop inter-Korean relations for national reconciliation and cooperation, and unwavering peace and co-prosperity, and to make efforts to realize through policy measures the aspiration and hope of all Koreans that the current developments in inter-Korean relations will lead to reunification.

The two leaders held candid and in-depth discussions on various issues and practical steps to advance inter-Korean relations to a new and higher dimension by fully implementing the Panmunjom Declaration, shared the view that the Pyongyang Summit will be an important historic milestone, and declared as follows.

1. The two sides agreed to expand the cessation of military hostilities in regions of confrontation such as the DMZ to the substantial removal of the danger of war across the entire Korean Peninsula and a fundamental resolution of the hostile relations.

2. 평양공동선언

• 2018년 9월 18일~20일, 평양

대한민국 문재인 대통령과 조선민주주의인민공화국 김정은 국무위원장은 2018년 9월 18일부터 20일까지 평양에서 남북정상회담을 진행하였다.

양 정상은 역사적인 판문점선언 이후 남북 당국간 긴밀한 대화와 소통, 다방면적 민간교류와 협력이 진행되고, 군사적 긴장완화를 위한 획기적인 조치들이 취해지는 등 훌륭한 성과들이 있었다고 평가하였다.

양 정상은 민족자주와 민족자결의 원칙을 재확인하고, 남북관계를 민족적 화해와 협력, 확고한 평화와 공동번영을 위해 일관되고 지속적으로 발전시켜 나가기로 하였으며, 현재의 남북관계 발전을 통일로 이어갈 것을 바라는 온 겨레의 지향과 여망을 정책적으로 실현하기 위하여 노력해 나가기로 하였다.

양 정상은 판문점선언을 철저히 이행하여 남북관계를 새로운 높은 단계로 진전시켜 나가기 위한 제반 문제들과 실천적 대책들을 허심탄회하고 심도 있게 논의하였으며, 이번 평양정상회담이 중요한 역사적 전기가 될 것이라는 데 인식을 같이 하고 다음과 같이 선언하였다.

1. 남과 북은 비무장지대를 비롯한 대치지역에서의 군사적 적대관계 종식을 한반도 전지역에서의 실질적인 전쟁위험 제거와 근본적인 적대관계 해소로 이어나가기로 하였다.

① The two sides agreed to fully abide by and faithfully implement the "Agreement on the Implementation of the Historic Panmunjom Declaration in the Military Domain" adopted as an annex to the Pyongyang Joint Declaration, and to actively take practical measures to transform the Korean Peninsula into a zone of permanent peace.
② The two sides agreed to engage in constant communication and close consultations to review the implementation of the Agreement and prevent accidental military clashes by promptly activating the Inter-Korean Joint Military Committee.

2. The two sides agreed to explore substantial measures to further advance exchanges and cooperation based on the spirit of mutual benefit and shared prosperity, and to develop the nation's economy in a balanced manner.
① The two sides agreed to hold a ground-breaking ceremony within this year for the connection of railways and roads along the east and west coasts.
② The two sides agreed, as conditions mature, to first normalize the Gaeseong industrial complex and the Mt. Geumgang Tourism Project, and to discuss the issue of forming a west coast joint special economic zone and an east coast joint special tourism zone.
③ The two sides agreed to actively engage in inter-Korean environment cooperation in order to protect and restore the natural ecosystem, and as a first step, to endeavor to achieve substantial results in the currently on-going forestry cooperation.
④ The two sides agreed to strengthen cooperation in the areas of prevention of epidemics, public health and medical care, including emergency measures to prevent the influx and spread of contagious diseases.

① 남과 북은 이번 평양정상회담을 계기로 체결한 '판문점선언 군사분야 이행합의서'를 평양공동선언의 부속합의서로 채택하고 이를 철저히 준수하고 성실히 이행하며, 한반도를 항구적인 평화지대로 만들기 위한 실천적 조치들을 적극 취해 나가기로 하였다.
② 남과 북은 남북군사공동위원회를 조속히 가동하여 군사분야 합의서의 이행실태를 점검하고 우발적 무력충돌 방지를 위한 상시적 소통과 긴밀한 협의를 진행하기로 하였다.

2. 남과 북은 상호호혜와 공리공영의 바탕 위에서 교류와 협력을 더욱 증대시키고, 민족경제를 균형적으로 발전시키기 위한 실질적인 대책들을 강구해 나가기로 하였다.
① 남과 북은 금년내 동, 서해선 철도 및 도로 연결을 위한 착공식을 갖기로 하였다.

② 남과 북은 조건이 마련되는 데 따라 개성공단과 금강산관광 사업을 우선 정상화하고, 서해경제공동특구 및 동해관광공동특구를 조성하는 문제를 협의해 나가기로 하였다.
③ 남과 북은 자연생태계의 보호 및 복원을 위한 남북환경협력을 적극 추진하기로 하였으며, 우선적으로 현재 진행 중인 산림분야 협력의 실천적 성과를 위해 노력하기로 하였다.
④ 남과 북은 전염성 질병의 유입 및 확산 방지를 위한 긴급조치를 비롯한 방역 및 보건·의료 분야의 협력을 강화하기로 하였다.

3. The two sides agreed to strengthen humanitarian cooperation to fundamentally resolve the issue of separated families.
① The two sides agreed to open a permanent facility for family reunion meetings in the Mt. Geumgang area at an early date, and to promptly restore the facility toward this end.
② The two sides agreed to give priority to resolving the issue of video reunions and exchange of video messages between the separated families through Inter-Korean Red Cross talks.

4. The two sides agreed to actively promote exchanges and cooperation in various fields so as to rejuvenate the sense of reconciliation and unity and to demonstrate the spirit of the Korean nation both internally and externally.
① The two sides agreed to further promote cultural and artistic exchanges, and stage a performance of the Pyongyang Art Troupe in Seoul in October this year as a first step.
② The two sides agreed to actively participate together in the 2020 Summer Olympic Games as well as other international games, and to cooperate in bidding for the joint hosting of the 2032 Summer Olympic Games.
③ The two sides agreed to hold meaningful events to celebrate the 11th anniversary of the October 4 Declaration, to jointly commemorate the 100th anniversary of the March First Independence Movement Day, and to hold working-level consultations toward this end.

5. The two sides shared the view that the Korean Peninsula must be turned into a land of peace free from nuclear weapons and nuclear threats, and that substantial progress toward this end must be made

3. 남과 북은 이산가족 문제를 근본적으로 해결하기 위한 인도적 협력을 더욱 강화해 나가기로 하였다.
① 남과 북은 금강산 지역의 이산가족 상설면회소를 빠른 시일내 개소하기로 하였으며, 이를 위해 면회소 시설을 조속히 복구하기로 하였다.
② 남과 북은 적십자 회담을 통해 이산가족의 화상상봉과 영상편지 교환 문제를 우선적으로 해결해 나가기로 하였다.

4. 남과 북은 화해와 단합의 분위기를 고조시키고 우리 민족의 기개를 내외에 과시하기 위해 다양한 분야의 협력과 교류를 적극 추진하기로 하였다.

① 남과 북은 문화 및 예술분야의 교류를 더욱 증진시켜 나가기로 하였으며, 우선적으로 10월 중에 평양예술단의 서울공연을 진행하기로 하였다.
② 남과 북은 2020년 하계올림픽경기대회를 비롯한 국제경기들에 공동으로 적극 진출하며, 2032년 하계올림픽의 남북공동개최를 유치하는 데 협력하기로 하였다.
③ 남과 북은 10 · 4 선언 11주년을 뜻 깊게 기념하기 위한 행사들을 의의 있게 개최하며, 3 · 1운동 100주년을 남북이 공동으로 기념하기로 하고, 그를 위한 실무적인 방안을 협의해 나가기로 하였다.

5. 남과 북은 한반도를 핵무기와 핵위협이 없는 평화의 터전으로 만들어 나가야 하며 이를 위해 필요한 실질적인 진전을 조속히 이루어 나가야 한다는 데 인식을

in a prompt manner.

① First, the North will permanently dismantle the Dongchang-ri missile engine test site and launch platform under the observation of experts from relevant countries.

② The North expressed its willingness to continue to take additional measures, such as the permanent dismantlement of the nuclear facilities in Yongbyon, as the United States takes corresponding measures in accordance with the spirit of the June 12 US-DPRK Joint Statement.

③ The two sides agreed to cooperate closely in the process of pursuing complete denuclearization of the Korean Peninsula.

6. Chairman Kim Jong Un agreed to visit Seoul at an early date at the invitation of President Moon Jae-in.

September 19, 2018

Moon Jae-in
President
Republic of Korea

Kim Jong Un
Chairman
State Affairs Commission
Democratic People's
Republic of Korea

같이 하였다.

① 북측은 동창리 엔진시험장과 미사일 발사대를 유관국 전문가들의 참관 하에 우선 영구적으로 폐기하기로 하였다.

② 북측은 미국이 6 · 12 북미공동성명의 정신에 따라 상응조치를 취하면 영변 핵시설의 영구적 폐기와 같은 추가적인 조치를 계속 취해 나갈 용의가 있음을 표명하였다.

③ 남과 북은 한반도의 완전한 비핵화를 추진해 나가는 과정에서 함께 긴밀히 협력해 나가기로 하였다.

6. 김정은 국무위원장은 문재인 대통령의 초청에 따라 가까운 시일 내로 서울을 방문하기로 하였다.

2018년 9월 19일

대 한 민 국	조선민주주의인민공화국
대 통 령	국 무 위 원 장
문 재 인	김 정 은

3. Agreement on the Implementation of the Historic Panmunjom Declaration in the Military Domain

Based on the common understanding that easing military tension and building confidence on the Korean Peninsula is integral to securing lasting and stable peace. South and North Korea reached a comprehensive agreement on the following with a view to fully implementing the historic 'Panmunjom Declaration for Peace, Prosperity and Unification of the Korean Peninsula' in the military domain.

1. South and North Korea agreed to completely cease all hostile acts against each other in every domain, including land, air and sea that are the source of military tension and conflict.

① The two sides discussed various measures to prevent armed conflict in every domain, including land, air and sea.

The two sides agreed to consult and resolve all matters that may lead to military conflict in a peaceful way and preclude the use of military force under any circumstance.

The two sides agreed to refrain from any action to infiltrate, attack or occupy each other' s area of jurisdiction by any means or method.

The two sides agreed to have consultations on matters including large-scale military exercises and military buildup aimed at each other, various forms of blockade, interdiction and obstruction of navigation as well as reconnaissance activities against each other through the 'Inter-Korean Joint Military Committee' .

The two sides agreed to continue to have consultations on various implementation measures related to the 'Panmunjom Declaration' ,

3. 역사적인 판문점선언 이행을 위한 군사분야합의서

• 2018년 9월 19일
Comprehensive Military Agreement, September 19, 2018

남과 북은 한반도에서 군사적 긴장 상태를 완화하고 신뢰를 구축하는 것이 항구적이며 공고한 평화를 보장하는 데 필수적이라는 공통된 인식으로부터 한반도의 평화와 번영, 통일을 위한 판문점선언을 군사적으로 철저히 이행하기 위하여 다음과 같이 포괄적으로 합의하였다.

1. 남과 북은 지상과 해상, 공중을 비롯한 모든 공간에서 군사적 긴장과 충돌의 근원으로 되는 상대방에 대한 일체의 적대행위를 전면 중지하기로 하였다.
   ① 쌍방은 지상과 해상, 공중을 비롯한 모든 공간에서 무력 충돌을 방지하기 위해 다양한 대책을 강구하였다.
   쌍방은 군사적 충돌을 야기할 수 있는 모든 문제를 평화적 방법으로 협의 · 해결하며, 어떤 경우에도 무력을 사용하지 않기로 하였다.
   쌍방은 어떠한 수단과 방법으로도 상대방의 관할구역을 침입 또는 공격하거나 점령하는 행위를 하지 않기로 하였다.
   쌍방은 상대방을 겨냥한 대규모 군사훈련 및 무력증강 문제, 다양한 형태의 봉쇄 차단 및 항행방해 문제, 상대방에 대한 정찰행위 중지 문제 등에 대해 '남북군사공동위원회' 를 가동하여 협의해 나가기로 하였다.
   쌍방은 군사적 긴장 해소 및 신뢰구축에 따라 단계적

in which both sides agreed to realize phased arms reduction, in accordance with alleviation of military tension and confidence building.

② The two sides agreed to cease various military exercises aimed at each other along the Military Demarcation Line from November 1, 2018.

On ground, the two sides agreed to cease all live-fire artillery drills and field training exercises at the regiment level and above within 5km from the MDL.

At sea, the two sides agreed to cease all live-fire and maritime maneuver exercises within the zone north of Deokjeok-do and south of Cho-do in the West Sea, and within the zone north of Sokcho and south of Tongcheon in the East Sea. The two sides also agreed to install covers on the barrels of coastal artilleries and ship guns and close all gunports within the zones.

In the air, the two sides agreed to ban tactical live-fire drills involving fixed-wing aircraft, including the firing of air-to-ground guided weapons within the designated No Fly Zones in the eastern and western regions of the MDL.

③ The two sides agreed to designate No Fly Zones for all aircraft types above the MDL, effective from 1 November, 2018. in the following way :

For fixed-wing aircraft. No Fly Zones will be designated within 40km from the MDL in the East (between MDL Markers No. 0646 and 1292) and within 20km from the MDL in the West (between MDL Markers No. 0001 and 0646).

Other No Fly Zones will be designated in the following way : for rotary-wing aircraft, within 10km from the MDL; for UAVs, within 15km from the MDL in the East and 10km from the MDL in the

군축을 실현해 나가기로 합의한 판문점선언을 구현하기 위해 이와 관련된 다양한 실행 대책들을 계속 협의하기로 하였다.

② 쌍방은 2018년 11월 1일부터 군사분계선 일대에서 상대방을 겨냥한 각종 군사연습을 중지하기로 하였다.

지상에서는 군사분계선으로부터 5km 안에서 포병 사격훈련 및 연대급 이상 야외기동훈련을 전면 중지하기로 하였다.

해상에서는 서해 남측 덕적도 이북으로부터 북측 초도 이남까지의 수역, 동해 남측 속초 이북으로부터 북측 통천 이남까지의 수역에서 포사격 및 해상 기동훈련을 중지하고 해안포와 함포의 포구 포신 덮개 설치 및 포문폐쇄 조치를 취하기로 하였다.

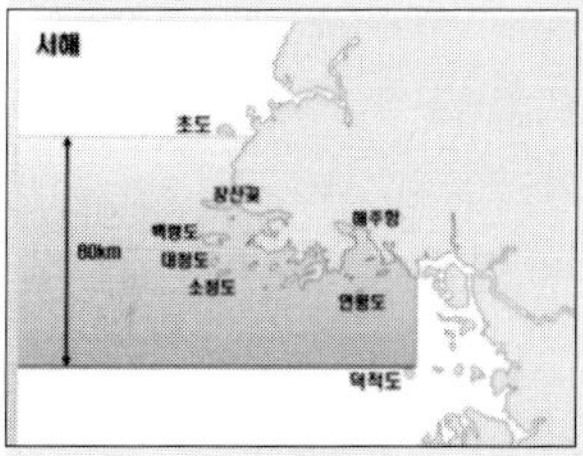

공중에서는 군사분계선 동 서부 지역 상공에 설정된 비행 금지구역 내에서 고정익항공기의 공대지유도무기사격 등 실탄사격을 동반한 전술훈련을 금지하기로 하였다.

③ 쌍방은 2018년 11월 1일부터 군사분계선 상공에서 모든 기종들의 비행금지구역을 다음과 같이 설정하기로 하였다.

고정익항공기는 군사분계선으로부터 동부지역(군사분계선표식물 제0646호부터 제1292호 까지의 구간)은 40km, 서부지역(군사분계선표식물 제0001호부터 제0646호까지의 구간)은 20km를 적용하여 비행금지구역을 설정한다.

회전익항공기는 군사분계선으로부터 10km로, 무인기는 동부지역에서 15km, 서부지역에서 10km로, 기

West; for hot-air balloons, within 25km from the MDL.

However, when the employment of aircraft becomes necessary such as in the cases involving fire-fighting, ground & maritime rescue, medical evacuation, weather observation and farming support, aircraft will be permitted to fly subject to prior notification to the other side. The No Fly Zones mentioned above do not apply to commercial aircraft (including cargo planes).

④ The two sides agreed to take measures to prevent any accidental military clash at all times in every domain, including land, air and sea.

To this end, the two sides agreed to apply a five-step procedure (Initial warning broadcast → Secondary warning broadcast → Initial warning fire → Secondary warning fire → Military action) on ground and at sea, and a 4-step procedure (Warning radio & signal → Interdiction flight → Warning fire → Military action) in the air.

The two sides agreed to implement the above revised procedures from November 1, 2018.

⑤ The two sides agreed to solve all military issues through peaceful consultations by maintaining permanent communication channels in order to prevent at all times any accidental military clash in every domain, including land, air and sea and by immediately notifying each other when an abnormal situation arises.

2. South and North Korea agreed to devise substantive military measures to transform the Demilitarized Zone into a peace zone.

① The two sides agreed to completely withdraw all Guard Posts (GP) that lie within 1km of each other as a preliminary measure to withdrawing all GPs within the DMZ. [Annex 1]

② The two sides agreed to demilitarize the Joint Security Area.

구는 25km로 적용한다.
다만, 산불 진화, 지 해상 조난 구조, 환자 후송, 기상 관측, 영농지원 등으로 비행기 운용이 필요한 경우에는 상대측에 사전 통보하고 비행할 수 있도록 한다. 민간 여객기(화물기 포함)에 대해서는 상기 비행금지 구역을 적용하지 않는다.

④ 쌍방은 지상과 해상, 공중을 비롯한 모든 공간에서 어떠한 경우에도 우발적인 무력충돌 상황이 발생하지 않도록 대책을 취하기로 하였다.
이를 위해 지상과 해상에서는 경고방송 → 2차 경고방송 → 경고사격 → 2차 경고사격 → 군사적 조치의 5개 단계로, 공중에서는 경교신 및 신호 → 차단비행 → 경고사격 → 군사적 조치의 4개 단계의 절차를 적용하기로 하였다.
쌍방은 수정된 절차를 2018년 11월 1일부터 시행하기로 하였다.
⑤ 쌍방은 지상과 해상, 공중을 비롯한 모든 공간에서 어떠한 경우에도 우발적 충돌이 발생하지 않도록 상시 연락체계를 가동하며, 비정상적인 상황이 발생하는 경우 즉시 통보하는 등 모든 군사적 문제를 평화적으로 협의하여 해결하기로 하였다.

2. 남과 북은 비무장지대를 평화지대로 만들어 나가기 위한 실질적인 군사적 대책을 강구하기로 하였다.
① 쌍방은 비무장지대 안에 감시초소(GP)를 전부 철수하기 위한 시범적 조치로 상호 1km 이내 근접해 있는 남북 감시초소들을 완전히 철수하기로 하였다.
② 쌍방은 판문점 공동경비구역을 비무장화하기로 하였다.

③ The two sides agreed to proceed with a pilot project of an Inter-Korean Joint Operation to Recover Remains within the DMZ.

④ The two sides agreed to continue consultations on military assurance measures related to joint survey and excavation of historical remains within the DMZ.

3. South and North Korea agreed to take military measures to prevent accidental military clashes and ensure safe fishing activities by turning the area around the Northern Limit Line in the West Sea into a maritime peace zone.

① The two sides reaffirmed the agreement related to the "prevention of accidental military clashes in the West Sea", signed during the 2nd Inter-Korean General-level Military Talks on 4 June, 2004 and agreed to fully restore and implement it.

② The two sides agreed to establish a maritime peace zone and a pilot joint fishing zone in the West Sea.

③ The two sides agreed to fully guarantee the safety of personnel and vessels that enter the maritime peace zone and the pilot joint fishing zone.

④ The two sides agreed to devise and implement inter-Korean joint patrol measures in order to deny illegal fishing and to ensure safe fishing activities for South and North Korean fishermen in the maritime peace zone and the pilot joint fishing zone.

4. South and North Korea agreed to devise military assurance measures necessary for invigorating exchanges, cooperation, contacts and visits.

① The two sides agreed to establish military measures to ensure passage, communication and customs in the East and West Transportation Corridors.

② The two sides agreed to devise military assurance measures to

③ 쌍방은 비무장지대내에서 시범적 남북공동유해발굴을 진행하기로 하였다.

④ 쌍방은 비무장지대 안의 역사유적에 대한 공동조사 및 발굴과 관련한 군사적 보장대책을 계속 협의하기로 하였다.

3. 남과 북은 서해 북방한계선 일대를 평화수역으로 만들어 우발적인 군사적 충돌을 방지하고 안전한 어로활동을 보장하기 위한 군사적 대책을 취해 나가기로 하였다.

① 쌍방은 2004년 6월 4일 제2차 남북장성급군사회담에서 서명한 '서해 해상에서의 우발적 충돌 방지' 관련 합의를 재확인하고, 전면적으로 복원 이행해 나가기로 하였다.

② 쌍방은 서해 해상에서 평화수역과 시범적 공동어로구역을 설정하기로 하였다.

③ 쌍방은 평화수역과 시범적 공동어로구역에 출입하는 인원 및 선박에 대한 안전을 철저히 보장하기로 하였다.

④ 쌍방은 평화수역과 시범적 공동어로구역 내에서 불법어로 차단 및 남북 어민들의 안전한 어로활동 보장을 위하여 남북 공동순찰 방안을 마련하여 시행하기로 하였다.

4. 남과 북은 교류협력 및 접촉 왕래 활성화에 필요한 군사적 보장대책을 강구하기로 하였다.

① 쌍방은 남북관리구역에서의 통행 통신 통관(3통)을 군사적으로 보장하기 위한 대책을 마련하기로 하였다.

② 쌍방은 동 서해선 철도 도로 연결과 현대화를 위한

connect and modernize eastern / western railways and roads.

③ The two sides agreed to establish a plan regarding issues related to permitting the use of Haeju Passage and Jeju Strait for North Korean vessels through consultations at the Inter-Korean Joint Military Committee.

④ The two sides agreed to devise military assurance measures for the shared use of Han River (Imjin River) Estuary.

5. South and North Korea agreed to devise various measures for mutual military confidence building.

① The two sides agreed to continue consultations regarding the installation and operation of direct communication lines between the respective military officials.

② The two sides agreed to have detailed consultations to resolve the issues concerning the composition and operation of the Inter-Korean Joint Military Committee.

③ The two sides agreed to fully implement all agreements reached between inter-Korean military authorities, and to regularly examine and assess the implementation progress.

6. This Agreement is effective from the date of exchange of the signed documents, following the procedures required for ratification from each party.

① Revisions and additions to the Agreement are permitted subject to agreement between both parties.

② Two copies of the Agreement have been made with equal effect.

19 September 2018

Republic of Korea	Democratic People' s Republic of Korea
Minister of National Defense	Minister of People' s Armed Forces
Song, Young Moo	Korean People' s Army General No Kwang-chol

군사적 보장대책을 강구하기로 하였다.

③ 쌍방은 북측 선박들의 해주직항로 이용과 제주해협 통과 문제 등을 남북군사공동위에서 협의하여 대책을 마련하기로 하였다.

④ 쌍방은 한강(임진강) 하구 공동이용을 위한 군사적 보장 대책을 강구하기로 하였다.

5. 남과 북은 상호 군사적 신뢰구축을 위한 다양한 조치들을 강구해 나가기로 하였다.

① 쌍방은 남북군사당국자 사이에 직통전화 설치 및 운영 문제를 계속 협의해 나가기로 하였다.

② 쌍방은 남북군사공동위원회 구성 및 운영과 관련한 문제를 구체적으로 협의 · 해결해 나가기로 하였다.

③ 쌍방은 남북군사당국간 채택한 모든 합의들을 철저히 이행하며, 그 이행상태를 정기적으로 점검 평가해 나가기로 하였다.

6. 이 합의서는 쌍방이 서명하고 각기 발효에 필요한 절차를 거쳐 그 문본을 교환한 날부터 효력을 발생한다.

① 합의서는 쌍방의 합의에 따라 수정 및 보충할 수 있다.

② 합의서는 2부 작성되었으며, 같은 효력을 가진다.

2018년 9월 19일

대 한 민 국	조선민주주의인민공화국
국 방 부 장 관	인 민 무 력 상
송 영 무	조선인민군 대장 노광철

[Annex 1]

Withdrawal of Guard Posts (GP) within the DMZ

1. The agreement is reached to completely withdraw all guard posts within the DMZ according to the following phases:

① Phase 1 : Withdrawal of all firearms and equipment

② Phase 2 : Withdrawal of all personnel on duty

③ Phase 3 : Complete destruction of all facilities

④ Phase 4 : Mutual verification

2. Various measures for the withdrawal of all guard posts from the DMZ will be implemented through close mutual coordination.

3. Historical landmarks, remains and artifacts that lie within the DMZ will be restored to their original state and be managed as joint assets of the Korean people.

4. The 11 guard posts that lie within 1km of each other in the DMZ from each side will be withdrawn as a preliminary measure.

① The 11 guard posts from each side that will be withdrawn are as follows :

- Eastern region
  - GPs from each side in between MDL Markers No. 1273 and 1278
  - GPs from each side in between MDL Markers No. 1123 and 1135
  - GPs from each side in between MDL Markers No. 0799 and 0808
- Midlands
  - GPs from each side in between MDL Markers No. 0652 and

[붙임 1]-2항의 ①의 감시초소 철수 세부사항인 [붙임1]

비무장지대내 상호 감시초소(GP) 철수

1. 비무장지대내 모든 감시초소를 완전히 철수하기로 합의하고, 다음과 같은 단계로 나누어 진행한다.
   ① 1단계 : 모든 화기 및 장비 철수
   ② 2단계 : 근무인원 철수
   ③ 3단계 : 시설물 완전파괴
   ④ 4단계 : 상호 검증
2. 비무장지대내 모든 감시초소 철수를 위한 제반 조치들은 상호 긴밀히 협조하여 진행한다.
3. 비무장지대 안에서 역사적 의미를 가지는 사적물과 유적, 유물들은 민족공동의 재산으로 원래 상태를 보존하여 관리한다.
4. 비무장지대 안의 1km 거리내에 근접해 있는 양측의 11개 감시 초소를 시범적으로 철수한다.
   ① 지역별로 철수할 양측 감시초소들은 다음과 같다.

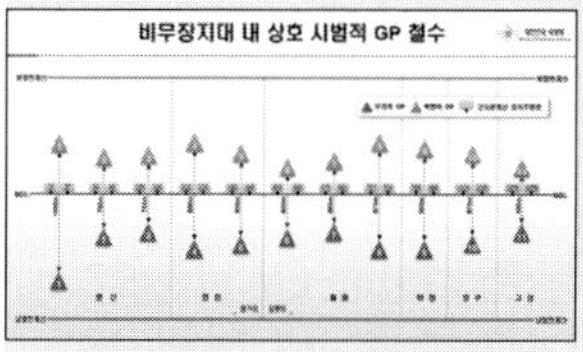

- 동부지역
  - 군사분계선표식물 제 1273호와 제 1278호 사이의 쌍방초소
  - 군사분계선표식물 제 1123호와 제 1135호 사이의 쌍방초소
  - 군사분계선표식물 제 0799호와 제 0808호 사이의 쌍방초소
- 중부지역
  - 군사분계선표식물 제 0652호와 제 0660호 사이

0660

- GPs from each side in between MDL Markers No. 0679 and 0683
- GPs from each side in between MDL Markers No. 0717 and 0724

\- Western region

- GPs from each side in between MDL Markers No. 0023 and 0027
- GPs from each side in between MDL Markers No. 0034 and 0043
- GPs from each side in between MDL Markers No. 0155 and 0166
- GPs from each side in between MDL Markers No. 0212 and 0216
- GPs from each side in between MDL Markers No. 0233 and 0240

② Withdrawal of the above 11 guard posts will be completed by December 31, 2018.

의 쌍방초소

- 군사분계선표식물 제 0679호와 제 0683호 사이의 쌍방초소
- 군사분계선표식물 제 0717호와 제 0724호 사이의 쌍방초소

- 서부지역

- 군사분계선표식물 제 0023호와 제 0027호 사이의 쌍방초소
- 군사분계선표식물 제 0034호와 제 0043호 사이의 쌍방초소
- 군사분계선표식물 제 0155호와 제 0166호 사이의 쌍방초소
- 군사분계선표식물 제 0212호와 제 0216호 사이의 쌍방초소
- 군사분계선표식물 제 0233호와 제 0240호 사이의 쌍방초소

② 양측의 11개 감시초소들에 대한 철수는 2018년 12월 31일까지 완료한다.

[Annex 2]

Demilitarization of the 'Joint Security Area in Panmunjom'

1. As a first step, a trilateral consultative body between South Korea, North Korea and the United Nations Command (UNC) will be established to consult and implement measures to demilitarize the Joint Security Area in Panmunjom.

① South, North and UNC (hereafter referred to as "The three parties") will remove all mines in the Joint Security Area in Panmunjom within 20 days, beginning on October 1, 2018.

② The three parties will completely withdraw guard posts, personnel and firearms within 5 days after the minesweeping is completed in the Joint Security Area in Panmunjom.

③ The three parties will withdraw unnecessary surveillance equipment from the Joint Security Area in Panmunjom, install any necessary surveillance equipment through consultations, and share related information with each other.

④ The three parties will jointly verify the completion of the measures to demilitarize the Joint Security Area in Panmunjom over 2 days.

⑤ The three parties will consult and decide on various issues such as the composition, mission and operation method of a joint administrative body that will he established after the demilitarization of the Joint Security Area in Panmunjom.

2. The Joint Security Area in Panmunjom will be managed after demilitarization in the following way :

① Establishment of guard duty personnel and guard posts

[붙임 2]—2항의 ② 공동경비구역 비무장화 세부사항인 **【붙임 2】**

'판문점 공동경비구역' 의 비무장화

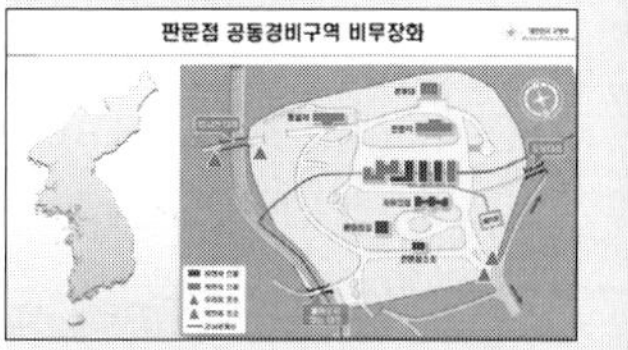

1. 첫 단계로 남 · 북 · 유엔사 3자 협의체를 구성하여 판문점 공동경비구역의 비무장화 조치 방안을 협의 · 이행한다.

① 남 · 북 · 유엔사(이하 3자로 표기)는 2018년 10월 1일부터 판문점 공동경비구역내 지뢰를 20일 안에 제거한다.
② 3자는 판문점 공동경비구역의 지뢰제거가 완료된 때로부터 5일 내에 쌍방초소들과 인원 및 화력장비를 전부 철수한다.
③ 3자는 판문점 공동경비구역 안에 불필요한 감시장비를 철수하고, 협의를 통해 필요한 감시장비를 추가 설치하며, 관련 정보를 상호 공유한다.

④ 3자는 판문점 공동경비구역의 비무장화 조치 완료 상태를 2일간 공동으로 검증한다.

⑤ 3자는 판문점 공동경비구역 비무장화 조치 완료 이후의 공동관리기구 구성 및 임무, 공동관리기구 운영방식 등과 관련한 제반 사항들을 협의하여 결정한다.

2. 비무장화 조치 이후 판문점 공동경비구역 관리운영은 다음과 같이 한다.
① 경비근무조직 및 초소 설치

- Guard duty will be carried out by up to 35 unarmed personnel from each side.
- The shift work of the personnel and issues related to patrol will be decided by each side unilaterally and notified to the other party.
- Guard duty personnel from both sides will wear a 15cm wide yellow armband with the words "Panmunjom Civil Police" written in blue on his/her left arm.
- A new Southern guard post will be installed at one end of the 'Panmunjom Bridge' in Northern Panmunjom; A new Northern guard post will be installed in the area near the entry check point of Southern Panmunjom. South and North Korean personnel will be on duty side-by-side.

② Building management
- Each side will manage the buildings on its side of the Joint Security Area in Panmunjom.
- In case a building in the Joint Security Area in Panmunjom requires repair or construction, such work will go ahead subject to approval by the Joint Administrative Body.

③ Visits
- Visits to the Joint Security Area in Panmunjom will be permitted between 0900 and 1700 hours.
- Freedom of movement is allowed for visitors and tourists within the Joint Security Area in Panmunjom.

- 경비근무는 양측이 각각 35명 이하의 비무장 인원으로 수행한다.
- 경비근무 인원의 교대 및 순찰과 관련한 사항은 쌍방이 각자 규정하고 상호 통보한다.

- 양측 경비근무 인원들은 노란색 바탕에 파란색으로 「판문점 민사경찰」이라고 쓴 너비 15cm의 완장을 왼팔에 착용한다.
- 공동경비구역 북측지역 「판문점 다리」 끝점에 남측 초소를, 남측지역 진입초소 일대에 북측 초소를 새로 설치하며, 남과 북이 근접하여 근무한다.

② 건물 관리
- 판문점 공동경비구역 내 자기측 건물들은 각측이 관리한다.
- 판문점 공동경비구역 내 건물을 보수 또는 새로 건설해야 할 경우, 공동관리기구의 승인하에 진행한다.

③ 참관(방문)
- 판문점 공동경비구역에 대한 참관은 09시부터 17시까지의 사이에 진행한다.
- 판문점 공동경비구역 내에서 관광객들과 참관 인원들의 자유왕래를 허용한다.

[Annex 3]

Pilot Inter-Korean Joint Remains Recovery Project within DMZ

1. Designation of Joint Remains Recovery Site and minesweeping

① Site designation for joint recovery of remains

- The site for the pilot joint remains recovery project is set in between MDL Markers No. 0489 and 0497 in Cheolwon, Gangwon Province, setting the MDL as the standard.
  The end coordinates for the site are as follows :
  - North : a) NL 38° 17′ 35″ EL 127° 05′ 22″
    b) NL 38° 18′ 23″ EL 127° 06′ 52″
  - South : c) NL 38° 16′ 38″ EL 127° 06′ 04″
    d) NL 38° 17′ 26″ EL 127° 07′ 33″
- All guard posts and obstacles within the joint remains recovery site must be withdrawn.

② Minesweeping

- All mines and explosives within the site for the pilot joint remains recovery project are to be completely removed between October 1 and November 30, 2018.
- The two sides will each carry out minesweeping operations from each end of the DMZ and work in the direction of the MDL.
- The two sides will conduct minesweeping operations for 4 hours each day from 1000 to 1200 hours and from 1500 to 1700 hours. The hours may be reduced or extended according to each circumstance.
- The two sides will install markers around the border of the mine-cleared area within the site designated for pilot joint remains recovery project and notify each other accordingly.

[붙임 3]–2항의 ③ 남북공동유해발굴 세부사항인 **【붙임 3】**

비무장지대 시범적 남북공동유해 발굴

1. 시범적 공동유해발굴지역 설정과 지뢰제거

① 시범적 공동유해발굴지역 설정

- 시범적 공동유해발굴지역은 강원도 철원지역 비무장지대의 군사분계선 표식물 제 0489호~제 0497호 구간에서 군사분계선을 기준으로 설정한다.

  시범적 공동유해발굴지역의 끝점좌표는 다음과 같다.

  • 북측 : ㄱ) 북위 38° 17′ 35″ 동경 127° 05′ 22″

  ㄴ) 북위 38° 18′ 23″ 동경 127° 06′ 52″

  • 남측 : ㄷ) 북위 38° 16′ 38″ 동경 127° 06′ 04″

  ㄹ) 북위 38° 17′ 26″ 동경 127° 07′ 33″

- 공동유해발굴지역에 있는 양측의 감시초소와 장애물들은 전부 철수한다.

② 지뢰제거

- 시범적 공동유해발굴지역의 지뢰와 폭발물은 2018년 10월 1일부터 11월 30일까지 완전히 제거한다.
- 쌍방은 비무장지대 자기측 경계선으로부터 군사분계선 방향으로 지뢰제거 작업을 진행한다.
- 쌍방은 지뢰제거 작업을 10시부터 12시까지, 15시부터 17시까지 총 4시간 진행하며, 경우에 따라 작업시간을 단축 또는 연장한다.

- 쌍방은 시범 공동유해발굴지역내 지뢰가 제거된 구역의 외곽선을 따라 일정한 간격으로 표식물을 설치하며, 그에 대하여 상대측에 통보한다.

- The use of the equipment and material necessary for minesweeping will be mutually coordinated.
- Any remains that are recovered during minesweeping operations will be exhumed and be jointly identified, consulted and processed.

2. Establishment of an inter-Korean road within the joint remains recovery site

① A 12m-wide road will be constructed between South and North Korea within the pilot joint remains recovery site in order to facilitate seamless progress in the joint efforts to recover the remains within the DMZ.

② Minesweeping will take place from each end of the DMZ towards the MDL prior to road construction. The road will be connected along the MDL.

③ The use of the material and equipment necessary for road construction, including excavators, will be mutually coordinated.

④ Personnel necessary for construction, quantity of equipment and its identification and the working hours are to be decided unilaterally and notified to each other.

⑤ If any personnel or vehicle needs to cross the MDL for the purpose of road construction, prior notification should be provided to the other side.

⑥ The construction of road must be completed by December 31, 2018.

3. Composition and operation of the Joint Remains Recovery Team

① Composition of the Joint Remains Recovery Team

- The two sides will establish a joint investigation team and a site command team, each with 5 members and headed by a colonel-level official.
- The Joint Remains Recovery Team will be comprised of personnel

- 지뢰제거에 필요한 장비와 자재 이용은 상호 협조한다.
- 지뢰제거 과정에서 발굴되는 유해들은 수습하여 공동으로 감식, 협의, 처리한다.

2. 공동유해발굴지역내 남북도로 개설

① 비무장지대 공동유해발굴의 원활한 추진을 위하여 시범적공동유해발굴지역내에 남북간 12m 폭의 도로를 개설한다.

② 도로는 각각 비무장지대 각측 경계선으로부터 군사분계선 방향으로 나가면서 지뢰제거 작업을 선행한 후 개설하며, 군사분계선상에서 연결한다.
③ 굴착기를 비롯하여 도로건설에 필요한 자재와 장비 이용에 대해 상호 협조한다.
④ 도로개설 작업 인원과 장비들의 수량과 식별표식, 작업시간은 각기 편리하게 정하며 이를 상대측에 통보한다.
⑤ 도로연결과 관련하여 일방의 인원이나 차량들이 군사분계선을 통과해야 할 경우에는 상대측에 사전에 통보한다.
⑥ 도로개설은 2018년 12월 31일까지 완료한다.

3. 공동유해발굴단 구성과 운영
① 공동유해발굴단의 구성
- 쌍방은 대령급을 책임자로 하여 각각 5명씩의 유해발굴 공동조사 및 현장지휘조를 구성한다.

- 유해발굴단은 쌍방에서 각각 80~100명 정도로 구

numbering 80 to 100 from each side.

- The two sides will complete the composition of the Joint Remains Recovery Team and notify each other by the end of February, 2019.

② Operation of the Joint Remains Recovery Team

- The joint investigation team and the site command team will jointly consult and resolve any practical issue that may arise with regards to the pilot joint remains recovery project within the DMZ.
- Joint remains recovery operations at the pilot site will take place from April 1 to October 31, 2019.
- In consideration of the seasonal climate, working hours for remains recovery at the pilot site will be from 0900 to 1200 hours and from 1500 to 1800 hours. If needed, the hours may be reduced or extended subject to agreement from both sides.

4. Safety assurance and joint management

① Any action that violates the personal safety of personnel from the other side is prohibited during the joint remains recovery process.

② Any material or equipment that may threaten personal safety such as weapons and explosives is not allowed into the joint remains recovery site.

③ Any action that provokes the other side in the joint remains recovery site is prohibited.

④ In case a natural disaster occurs within the joint remains recovery site, the two sides will establish damage repair plans and cooperate accordingly.

⑤ Once joint remains recovery is completed, each side will manage its own area and road between the MDL and its end point of the DMZ. Issues regarding the use of roads will be decided in the future through consultations.

성한다.

- 쌍방은 2019년 2월말까지 공동유해발굴단 구성을 완료하여 상호 통보한다.

② 공동유해발굴단의 운영

- 유해발굴 공동조사 및 현장지휘조는 비무장지대의 시범적 공동유해발굴과 관련하여 제기되는 실무적 문제들을 공동으로 협의하여 해결한다.
- 시범적 공동유해발굴지역에서의 공동유해발굴은 2019년 4월 1일부터 10월 31일까지의 기간동안 진행한다.
- 시범적 공동유해발굴지역에서의 유해발굴 작업시간은 계절 특성을 고려하여 9시부터 12시까지, 15시부터 18시까지로 하며, 필요한 경우 쌍방이 합의하여 단축 또는 연장한다.

4. 안전보장 및 공동관리

① 공동유해발굴 과정에서 상대측 인원들의 신변안전을 침해하는 일체 행위를 금지한다.

② 공동유해발굴지역에 무기, 폭발물 등 신변안전을 위협하는 물자 및 장비의 반입을 불허한다.

③ 공동유해발굴지역에서 상대측을 자극하는 행위를 금지한다.

④ 공동유해발굴지역에서 자연재해가 발생하는 경우, 피해복구 대책을 수립하여 상호 협력한다.

⑤ 공동유해발굴이 완료된 후 중앙군사분계선으로부터 자기측 경계선까지의 지역과 도로에 대해서는 각 측이 책임지고 관리하며, 도로이용과 관련한 문제는 추후에 협의하여 결정한다.

[Annex 4]

Preventing Accidental Military Clashes, Establishing a Maritime Peace Zone and Ensuring Safe Fishing Activities in the West Sea

1. Establishment of a maritime peace zone

1) Scope of the maritime peace zone

The scope of the maritime peace zone will be determined in consideration of the factors including geographic location of the islands under the jurisdiction of each side, density of passage of vessels and fixed sea routes. Specific boundary lines will be confirmed through consultations at the Inter-Korean Joint Military Committee.

2) Rules pertaining to entry into the maritime peace zone

① Only unarmed vessels are allowed entry into the maritime peace zone. If the entry of naval ships is unavoidable, it can enter the zone subject to prior notification to and subsequent approval of the other side.

② The number of vessels operating in the maritime peace zone will be decided by the two sides through consultations.

The plans for entry and operation of vessels within the maritime peace zone must be notified to the other side 48 hours prior to its entry.

③ Entry hours are as follows: from April to September, 0700 to 1800 hours; from October to March, 0800 to 1700 hours. The entry hours may be revised through mutual consultations.

3) Rules pertaining to activities within the maritime peace zone

[붙임 4]–3항의 ② 평화수역과 시범적 공동어로구역 세부사항인【붙임 4】

서해 해상에서 우발적 충돌방지 및 평화수역 설정, 안전한 어로활동 보장

1. 평화수역 설정

1) 평화수역 범위

평화수역 범위는 쌍방의 관할하에 있는 섬들의 지리적 위치, 선박들의 항해밀도, 고정항로 등을 고려하여 설정하되, 구체적인 경계선은 남북군사공동위에서 협의하여 확정하기로 하였다.

2) 평화수역 출입 질서

① 평화수역에는 쌍방의 비무장선박들만 출입한다. 해군 함정들이 평화수역으로 불가피하게 진입하여야 할 필요성이 제기되는 경우에는 상대측에 사전 통보하고 승인하에 출입한다.

② 평화수역 내에서 활동하는 선박 수는 쌍방이 협의하여 결정한다.

선박들의 평화수역 출입 및 활동 계획은 48시간 전에 상호 통보한다.

③ 평화수역 출입시간은 4월부터 9월까지는 7시부터 18시, 10월부터 3월까지는 8시부터 17시까지로 하며, 필요한 경우 상호 협의하여 변경할 수 있다.

3) 평화수역에서의 행동 질서

① Vessels from the South must not cross the Northern boundary line and vessels from the North must not cross the Southern boundary line. All activities are limited to those of peaceful nature.
Vessels that commit hostile acts in the other side' s waters, outside the peace zone, must be restrained immediately. After the matter is reported to the other side, the vessel will be dealt with following inter-Korean consultations.
② For the purpose of identification within the maritime peace zone, vessels from both sides must hoist a 900mm wide. 600mm long Korean Peninsula flag. Vessels from the South must hoist the flag on the left of the mast and vessels from the North on the right of the mast.
③ Any words and actions that may provoke the other side, including psychological warfare, are not allowed within the maritime peace zone.
④ If an accidental clash occurs between civilian vessels in the maritime peace zone, each side must immediately withdraw all of its vessels from the zone, resolve the matter through either inter-Korean military communication lines or inter-Korean working-level military talks, and establish meticulous plans to prevent any recurrence.

4) Humanitarian cooperation in the maritime peace zone
If individuals, vessels, naval ships and aircraft enter the maritime peace in force majeure conditions such as engine failure, distress or misnavigation resulting from deterioration of weather, each side must immediately notify the other through communication means at its disposal.
When such emergency situation arises within the maritime peace zone, necessary measures must be taken through mutual

① 평화수역에서 남측 선박은 평화수역 북경계선을, 북측선박은 평화수역 남경계선을 넘지 않으며, 평화적 활동으로만 한정한다.

평화수역을 벗어나 상대측 수역에서 상대측에 대한 적대행위를 하는 선박들에 대해서는, 적대행위를 즉시 제지시키고 상대측에 통보한 후 남과 북이 협의하여 처리한다.

② 평화수역에서 쌍방 선박들은 식별을 위하여 가로 900mm×세로 600mm 크기의 한반도기를 남측 선박은 마스트 좌현 기류줄에, 북측 선박은 마스트 우현 기류줄에 게양한다.

③ 평화수역에서 심리전행위를 비롯하여 상대측을 자극하는 일체 언행을 하지 않는다.

④ 평화수역에서 민간선박들 사이에 우발적인 충돌이 발생하는 경우 즉시 자기측의 모든 선박 등을 평화수역 밖으로 철수시키고, 남북군통신선 혹은 남북군사실무회담을 통하여 사태를 수습하며 재발방지 대책을 철저히 세운다.

4) 평화수역에서 인도주의적 협력 문제

개별 인원과 선박, 함정, 항공기 등이 기관고장, 조난, 기상악화로 인한 항로미실 등 불가피한 상황으로 평화수역에 들어가는 경우 상대측에 연락 수단을 통해 즉시 통보한다.

평화수역 안에서 상기와 같은 긴급한 상황이 발생할 경우 상호 협력하여 필요한 조치를 취한다.

cooperation.

5) Utilization of the maritime peace zone

The two sides agreed to continue to explore options for peaceful utilization of the zone, including marine survey, joint survey and passage of civilian vessels in accordance with the principle of easing military tension, confidence building as well as public management and interest.

2. Establishment of a pilot joint fishing zone

1) Scope of the pilot joint fishing zone

The pilot joint fishing zone will be established in between PY-do (South) and Jangsan-got (North). Specific boundary lines will he finalized through consultations at the Inter-Korean Joint Military Committee.

2) Operation of the pilot joint fishing zone

① Vessels that wish to operate within the pilot zone must submit an entry request form 2 days (48 hours) prior to the planned entry, including the name of organization, name of the captain (representative), crew list, vessel name, entry route and date of operation.

② Responsible agencies must notify the other side the result of its review of the request form 1 day (24 hours) prior to departure. If entry of a vessel is denied, the agency must also notify valid reasons for denial.

③ Responsible agencies may allow vessels to remain within the pilot fishing zone for up to 5 days, if the vessel had requested permission for consecutive fishing operations for a certain period.

④ Vessels entering the pilot joint fishing zone must use the mutually approved route and will be controlled by fishery guidance boats

5) 평화수역 활용

쌍방은 서해 평화수역을 군사적 긴장완화 신뢰구축 및 공영 공리 원칙에 맞게 해양측량 · 공동조사, 민간 선박 운항 등 평화적으로 활용하기 위한 방안을 계속 협의하기로 하였다.

2. 시범 공동어로구역 설정

1) 시범 공동어로구역 범위

시범 공동어로구역 범위는 남측 백령도와 북측 장산곶 사이에 설정하되, 구체적인 경계선은 남북군사공동위원회에서 협의하여 확정하기로 하였다.

2) 시범 공동어로구역 운용질서

① 시범 공동어로구역에서 조업하고자 하는 어선들은 기관(소속)명칭, 선장(대표자), 선원명단, 어선명, 출입경로, 조업일자 등이 기록된 출입신청 문건을 출입 예정 2일(48시간) 전까지 상대측에 제출한다.

② 쌍방 해당기관은 어선의 출항예정 1일(24시간) 전까지 출입 신청서 검토 결과(동의 여부)를 상대측에 통보하되, 출입이 허가되지 않는 선박에 대해서는 타당한 사유와 함께 통보한다.

③ 쌍방 해당기관은 시범 공동어로구역내에서 일정 기간 연속조업을 신청한 어선들에 대하여 최대 5일간까지 공동어로구역내 체류를 허가한다.

④ 시범 공동어로구역에 출입하는 어선들은 상호 승인한 경로를 이용하며 쌍방 어업지도선의 통제를 받

from each side.

⑤ If, in the future, the joint fishing zone is expanded within the maritime peace zone, entry regulations for South and North Korean fishing vessels will be implemented through mutual consultations.

3. Joint patrol to deny illegal fishing vessels and ensure safe fishing activities

1) Organization of the Inter-Korean Joint Patrol Team

① The two sides will establish an 'Inter-Korean Joint Patrol Team' composed of maritime coast guard boats (patrol boats). The joint patrol boats must be 250 tons or below.

② 3 joint patrol boats will be assigned to each side (total of 6), but the number may be adjusted subject to agreement.

③ The boats that belong to the Inter-Korean Joint Patrol Team must hoist a 900mm wide, 600mm long yellow flag on the top of the mast.

2) Mission of the Inter-Korean Joint Patrol Team

① The Inter-Korean Joint Patrol Team interdicts illegal fishing vessels from third countries that attempt to enter the maritime peace zone through the pilot joint fishing zone, and controls and deals with them through close coordination.

② The Inter-Korean Joint Patrol Team controls the order among fishing vessels from the South and the North as well as fishery guidance boats that enter into the pilot joint fishing zone.

③ The Inter-Korean Joint Patrol Team rescues vessels drifting due to engine failure, distress and deterioration of weather, and returns them in accordance with humanitarian principles.

3) Operation of the Inter-Korean Joint Patrol Team

① Patrol boats of the Inter-Korean Joint Patrol Team will be

도록 한다.

⑤ 향후 평화수역 내에 공동어로구역을 확대하여 설정하는 경우, 남북 어선들의 공동어로구역 출입질서와 관련한 사항은 상호 협의하여 시행한다.

3. 불법어선 차단 및 안전한 어로활동 보장을 위한 공동순찰

1) 남북공동순찰대 조직

① 쌍방은 해경정(경비정)으로 '남북공동순찰대' 를 조직한다. 공동순찰정은 250톤급 이하로 한다.

② 공동순찰을 실시하는 순찰정은 쌍방이 각각 3척(총 6척)으로 하며 합의에 따라 척수를 조정할 수 있다.

③ 남북공동순찰대의 순찰정들은 가로 900mm×세로 600mm 크기의 황색 깃발을 선박 마스트 상단 부분에 게양한다.

2) 남북공동순찰대의 임무

① 서해 해상의 시범 공동어로구역을 통하여 평화수역에 진입하는 제 3국 불법 어선들을 차단하며, 상호 긴밀하게 협조하여 단속 처리한다.

② 시범 공동어로구역에 출입하는 남북 어선들과 어업지도선들의 항행질서를 통제한다.

③ 기관고장, 조난, 기상악화 등으로 표류하는 쌍방의 선박들을 구조하고 인도주의적 원칙에 따라 돌려보낸다.

3) 남북공동순찰대의 운용

① 남북공동순찰대의 순찰정들은 공동어로구역 내로

prohibited from entering the joint fishing zone. However, in case of emergency such as distress and rescue of personnel within the fishing zone, the patrol boats may enter upon notification to the other side.

② Joint patrol will take place on a date agreed by both sides in consideration of the fishing schedule and interdiction of illegal fishing vessels from third countries.

③ Joint patrol will take place in principle during the day (April-September : 0800-1800 hours, October-March : 0900-1700 hours). Each side must notify its patrol team' s schedule to the other 24 hours in advance.

Any situation that arises during the night is to be resolved through consultations between the two sides.

④ Joint patrol route follows around the outer boundary of the joint fishing zone either clockwise or anti-clockwise, depending on mutual agreement.

⑤ The Inter-Korean Joint Patrol Team follows each of its own superior authority' s command. Communications and call signs between patrol boats will adhere to the 'June 4 Agreement' of 2004.

⑥ The two sides will refrain from any provocative comment or action during joint patrol. In the event of a contingency, the patrol boat must be immediately separated and the matter must be resolved through mutual consultations.

진입을 금지한다. 다만, 공동어로구역내 조난, 인명구조 등 긴급한 상황 발생시에는 상대측에 통보 후 진입할 수 있다.
② 공동순찰은 조업 일정과 제3국 불법조업선박 차단 등과 연계하여 필요에 따라 쌍방이 합의한 날짜에 진행한다.
③ 공동순찰은 주간(4월부터 9월까지는 8시부터 18시까지, 10월부터 3월까지는 9시부터 17시까지)에 진행하는 것을 원칙으로 하며 자기측 공동순찰대의 순찰계획은 24시간 전에 상대측에 통보한다.
야간에 발생하는 상황은 쌍방이 협의하여 처리한다.

④ 공동순찰은 공동어로구역 외곽선을 따라 순방향 또는 역방향으로 상호 합의하여 기동한다.

⑤ 남북공동순찰대는 각기 자기측 상부의 지휘에 따라 행동하며, 쌍방 순찰정간 교신, 호출부호 등은 2004년 6.4 합의서를 준용하여 적용한다.
⑥ 공동순찰 과정에서 상대측을 자극하는 발언이나 행동을 하지 않으며, 상황 발생시 즉시 순찰정을 격리시키고 상호 협의를 통해 문제를 해결한다.

[Annex 5]

Military Assurance for the Joint Use of Han (Imjin) River Estuary

1. Establishment of a joint utilization zone

① The zone within the Han (Imjin) River Estuary stretching 70km long; in the South, from the Eastern end point of the Gimpo peninsula to the south western end point of Gyodong-do, and in the North, from Kaesong-si Panmun-kun Imhan- ri to Hwanghaenam-do Yeonan-kun Haenam-ri will be designated as the joint utilization zone.

② All practical military issues arising from within the joint utilization zone will be dealt with through consultations between the two sides.

2. Joint survey

① Field survey of the joint utilization zone will be carried out jointly by the end of December, 2018.

② The joint survey team will be composed of 10 people from each side, including subject matter experts.

③ The use of equipment, hardware and vessels required for the joint survey will be subject to mutual cooperation.

④ Any comment or action that may provoke the other side is prohibited among the site survey crew. They may not carry any explosives, weapons or live rounds.

⑤ In case of an emergency such as a natural disaster during the joint survey, the team may anchor at a nearby location under the other side' s jurisdiction, and the safety and comfort of the team members must be ensured.

3. Military assurance measures within the joint utilization zone

① A list that includes the relevant information on the personnel and

[붙임 5]

한강(임진강) 하구 공동이용의 군사적 보장

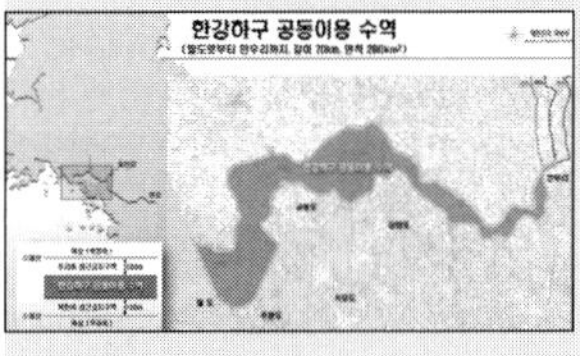

1. 공동이용수역 설정

① 남측의 김포반도 동북쪽 끝점으로부터 교동도 서남쪽 끝점까지, 북측의 개성시 판문군 임한리로부터 황해남도 연안군 해남리까지 70km에 이르는 한강(임진강) 하구 수역을 공동이용수역으로 설정한다.

② 공동이용수역 안에서 제기되는 모든 군사 실무적 문제들은 쌍방이 협의하여 처리한다.

2. 공동조사

① 공동이용수역에 대한 현장조사는 2018년 12월말까지 공동으로 진행한다.

② 공동조사단은 쌍방에서 해당 전문가를 포함하여 각각 10여 명의 인원들로 구성한다.

③ 공동조사에 필요한 장비, 기재, 선박 이용 문제는 상호 협조한다.

④ 현장조사 인원들은 상대측을 자극하는 발언이나 행동을 금지하며, 폭발물 및 각종 무기, 총탄 등을 휴대하지 않는다.

⑤ 공동조사중 자연재해 등 긴급상황이 발생하는 경우 가까운 상대측 지역에 정박할 수 있으며 상대측 인원들의 신변 안전과 편의를 보장한다.

3. 공동이용수역에서의 군사적 보장 대책

① 공동이용수역에 출입하는 인원 및 선박(선박의 형

vessel (type, length and weight, purpose of entry, size of crew, cargo on board) due to enter the joint utilization zone must be notified to the other side 1 day in advance via the Western inter-Korean military communication line.

② Check points for both sides in mutually agreed upon locations within the joint utilization zone will be established, where personnel and vessels will be inspected.

③ All vessels sailing through the joint utilization zone will not be allowed to approach within 100m of the other side' s boundary line.

④ In consideration of the seasonal influence on visual identification capability, passage hours for vessels in the joint utilization zone will be as follows: 0700 to 1900 hours from April 1 to September 30, and 0800 to 1800 hours from October 1 to March 31.

⑤ No personnel or vessel that sails through the joint utilization zone is allowed to carry surveillance and reconnaissance equipment, explosives, other weapons or live-rounds.

⑥ Any comment or action that may provoke the other side will be prohibited in the joint utilization zone.

⑦ Vessels from each side may not contact or communicate with vessels from the other side unless for the purpose of exchanging navigational signals to avoid mutual collision.

⑧ If a vessel or individual drifts within the joint utilization zone or an emergency situation arises due to other causes, both sides must cooperate under humanitarian principles.

4. Military assurance measures related to inter-Korean exchanges and cooperation within the joint utilization zone will he established through consultations between the two sides.

태, 길이 및 톤수, 출입목적, 탑승인원 수, 적재화물)에 대한 관련 내용을 작성하여 서해지구 남북 군통신선을 통해 1일 전에 상호 통보한다.

② 공동이용수역내 쌍방이 합의한 위치에 각측 통행검사소를 설치하고 출입인원과 선박들에 대한 통행검사를 진행한다.

③ 공동이용수역에 항행하는 선박들은 상대측 경계선으로부터 100m 이내에 접근하지 못한다.

④ 공동이용수역에서 선박들의 통행시간은 육안으로 관측이 가능한 계절적 특성을 고려하여 4월 1일부터 9월 30일까지는 7시부터 19시까지, 10월 1일부터 3월 31일까지는 8시부터 18시까지로 한다.

⑤ 공동이용수역에서 항행하는 인원 및 선박들은 정찰 및 감시 장비, 폭발물 및 각종 무기, 총탄 등을 일체 휴대하지 않는다.

⑥ 공동이용수역에서 상대측을 자극하는 발언이나 행동을 하지 않는다.

⑦ 쌍방 선박은 상호 충돌을 피하기 위한 항행신호 교환을 제외하고 상대측 선박과 연락 및 통신을 할 수 없다.

⑧ 공동이용수역 내에서 선박이나 인원이 표류하거나 기타 요인에 의하여 긴급한 상황이 발생하는 경우 인도주의 원칙에서 상호 협력한다.

4. 공동이용수역에서의 남북 교류협력 관련 군사적 보장대책은 쌍방 협의를 통해 마련해 나간다.

4. Major Inter-Korean Agreement List

1. The July 4 South-North Joint Communiqué

2. Agreement on Reconciliation, Non-aggression and Exchanges And Cooperation Between the South and the North

3. The June 15th South-North Joint Declatation

4. Declaration on the Advancement of South-North Korean Relations, Peace and Prosperity

5. Panmunjom Declaration on Peace, Prosperity and Reunification of the Korean Peninsula

7.4 남북공동성명/ The July 4 South-North Joint Communiqué
http://www.mofa.go.kr/www/brd/m_3984/view.do?seq=341001&srchFr=&srchTo=&srchWord=&srchTp=&multi_itm_seq=0&itm_seq_1=0&itm_seq_2=0&company_cd=&company_nm=&page=3

남북 사이 화해와 불가침 및 교류협력/ Agreement on Reconciliation, Non-aggression and Exchanges And Cooperation Between the South and the North
http://www.mofa.go.kr/www/brd/m_3984/view.do?seq=341000&srchFr=&srchTo=&srchWord=&srchTp=&multi_itm_seq=0&itm_seq_1=0&itm_seq_2=0&company_cd=&company_nm=

4. 주요 남북합의서 목록

1. 7.4남북공동성명

2. 남북 사이의 화해와 불가침 및 교류 협력에 관한 합의서 (남북기본합의서)

3. 6.15 남북공동선언

4. 남북관계 발전과 평화번영을 위한 선언 (10.4정상선언)

5. 한반도의 평화와 번영, 통일을 위한 판문점 선언 (판문점 선언)

6.15 남북공동선언/ Peace Agreements Digital Collection
http://theme.archives.go.kr/next/unikorea/six/six04.do

남북 관계발전과 평화번영/ Declaration on the Advancement of South-North Korean Relations, Peace and Prosperity
http://www.mofa.go.kr/www/brd/m_3984/view.do?seq=364829&srchFr=&srchTo=&srchWord=&srchTp=&multi_itm_seq=0&itm_seq_1=0&itm_seq_2=0&company_cd=&company_nm=&page=1

판문점 선언/ Panmunjom Declaration on Peace, Prosperity and Reunification of the Korean Peninsula
http://www.mofa.go.kr/www/brd/m_3976/view.do?seq=368007&srchFr=&srchTo=&srchWord=&srchTp=&multi_itm_seq=0&itm_seq_1=0&itm_seq_2=0&company_cd=&company_nm=&page=1

5. Joint Statement of President Trump and Chairman Kim at the Singapore Summit (June 12, 2018)

President Donald J. Trump of the United States of America and Chairman Kim Jong Un of the State Affairs Commission of the Democratic People' s Republic of Korea(DPRK) held a first, historic summit in Singapore on June 12, 2018.

President Trump and Chairman Kim Jong Un conducted a comprehensive, in-depth, and sincere exchange of opinions on the issues related to the establishment of new U.S.-DPRK relations and the building of a lasting and robust peace regime on the Korean Peninsula. President Trump committed to provide security guarantees to the DPRK, and Chairman Kim Jong Un reaffirmed his firm and unwavering commitment to complete denuclearization of the Korean Peninsula.

Convinced that the establishment of new U.S.-DPRK relations will contribute to the peace and prosperity of the Korean Peninsula and of the world, and recognizing that mutual confidence building can promote the denuclearization of the Korean Peninsula, President Trump and Chairman Kim Jong Un state the following:

1. The United States and the DPRK commit to establish new U.S.-DPRK relations in accordance with the desire of the peoples of the two countries for peace and prosperity.
2. The United States and the DPRK will join their efforts to build a lasting and stable peace regime on the Korean Peninsula.

5. 북미 싱가포르 정상회담
(2018년 6월 12일)

미합중국 도널드 J. 트럼프 대통령과 조선민주주의인민공화국 김정은 국무위원장간의 싱가포르 조미정상회담 공동합의문

미합중국 대통령 도널드 J. 트럼프와 조선민주주의인민공화국 국무위원장 김정은은 2018년 6월 12일 싱가포르에서 처음으로 역사적인 회담을 개최하였다.

트럼프 대통령과 김정은 국무위원장은 새로운 조미관계 수립과 한반도에서의 지속적이고 강건한 평화체제 건설에 관한 의제에 대하여 포괄적이고 면밀하며 진실성 있는 의견교환을 이뤄냈다. 트럼프 대통령은 조선민주주의인민공화국에 체제안전보장을 약속하였고, 김정은 국무위원장은 단호하고 확고하게 한반도에서의 비핵화를 완성해 나갈 것을 약속하였다.

새로운 조미관계의 수립이 한반도에서의 평화와 번영에 기여함을 확신하고 양국간 상호신뢰 구축이 한반도 비핵화를 촉진시킬 수 있음을 인지하면서 트럼프 대통령과 김정은 국무위원장은 다음과 같이 선언한다.

1. 미합중국과 조선민주주의인민공화국은 양국 국민들의 평화와 번영을 향한 염원에 부합하면서 새로운 조미관계를 수립하기로 약속하였다.
2. 미합중국과 조선민주주의인민공화국은 지속적이고 안정적인 평화체제를 한반도 내에서 구축하기 위한 노력에 협력하기로 하였다.

3. Reaffirming the April 27, 2018 Panmunjom Declaration, the DPRK commits to work toward complete denuclearization of the Korean Peninsula.
4. The United States and the DPRK commit to recovering POW/MIA remains, including the immediate repatriation of those already identified.

Having acknowledged that the U.S.-DPRK summit-the first in history-was an epochal event of great significance in overcoming decades of tensions and hostilities between the two countries and for the opening up of a new future, President Trump and Chairman Kim Jong Un commit to implement the stipulations in this joint statement fully and expeditiously. The United States and the DPRK commit to hold follow-on negotiations, led by the U.S. Secretary of State, Mike Pompeo, and a relevant high-level DPRK official, at the earliest possible date, to implement the outcomes of the U.S.-DPRK summit.

President Donald J. Trump of the United States of America and Chairman Kim Jong Un of the State Affairs Commission of the Democratic People's Republic of Korea have committed to cooperate for the development of new U.S.-DPRK relations and for the promotion of peace, prosperity, and security of the Korean Peninsula and of the world.

DONALD J. TRUMP

President of the United States of America

KIM JONG UN

Chairman of the State Affairs Commission of the Democratic People's Republic of Korea

June 12, 2018, Sentosa Island, Singapore

3. 2018년 4월 27일 발표된 판문점 선언의 구체적 실행을 재확인하며, 조선민주주의인민공화국은 한반도에서의 완전한 비핵화를 위해 노력하기로 약속하였다.
4. 미합중국과 조선민주주의인민공화국은 전쟁포로(POW; Prisoner of War)와 전시행방불명자(MIA; Missing in Action)에 대한 유해발굴과 신원 기확인자(이미 확인된 사람)에 대한 즉각적인 유해송환을 추진하기로 합의하였다.

역사상 최초로 개최된 조미정상회담이 지난 수십년 동안 양국간 긴장과 적대로 점철된 시간을 극복하고 다가올 새로운 미래를 준비하는 데 있어서 커다란 의미를 가지고 있는 신기원적 사건임을 인지하면서, 트럼프 대통령과 김정은 국무위원장은 이번 합의문에 규정된 사항들을 완전히 그리고 신속하게 이행할 것을 약속하였다. 미합중국과 조선민주주의인민공화국은 미합중국 국무장관인 마이크 폼페오와 이에 상응하는 조선민주주의인민공화국 유관 고위당국자의 협의를 통해 가능한 최대한 빠른 시일내에 조미정상회담의 결과물을 이행하기 위한 추가 협상을 진행하기로 약속하였다.

미합중국 대통령 도널드 J. 트럼프와 조선민주주의인민공화국 국무위원장 김정은은 새로운 조미관계의 발전과 한반도와 세계의 평화, 번영 그리고 안정의 촉진을 위해 협력할 것을 다짐하였다.

미합중국 대통령 도널드 J. 트럼프

조선민주주의인민공화국 국무위원회 위원장 김정은

2018년 6월 12일 센토사 섬, 싱가포르

참고문헌

1) 합참정보본부, 『군사정전위원회 편람』(제8집)

2) 이문항, 『JSA-판문점(1953∼1994)』(서울 : 소하, 2001)

3) 이상철, 『한반도 정전체제』(서울 : 한국국방연구원, 2012)

4) 손규석 · 조성훈 · 김상원, 『6.25 전쟁과 유엔군』(서울 : 국방부 군사편찬연구소, 2015)

5) 한국사 데이터베이스－국사편찬위원회(db.history.go.kr/id/pd_002_0180)

6) 국방부, 『북방한계선에 관한 우리의 입장』(서울 : 신오성기획인쇄사, 2007)

7) 정전협정전문

8) 한국전쟁과 관련된 유엔 안보리 결의

가) 전쟁행위 중지에 관한 유엔결의안 82호(1950년 6월 25일)

나) 대한민국에 대한 군사원조에 관한 유엔결의안 83호(1950년 6월 27일)

다) 국제연합 통합군사령부 설치에 관한 유엔결의안 84호(1950년 7월 7일)

라) 한국육 · 해 · 공군작전지휘권 이양에 관한 이승만대통령과 맥아더유엔군총사령관간의 교환공한(1950년 7월 14일)

마) 한국정전에 관한 유엔결의안(1950년 12월 14일)

바) 중공을 한국에서의 침략자로 규정하는 유엔결의안(1951년 2월 1일)

사) 한국포로에 관한 유엔결의안(1951년 2월 1일)

아) 휴전조인에 관한 이승만대통령의 성명(1953년 7월 27일)

자) 한국휴전에 관한 참전16개국 공동정책선언(1953년 7월 27일)

차) 한국참전 16개국의 제네바회담 공동성명(1954년 6월 15일)

9) 한반도 평화와 번영, 통일을 위한 판문점 선언(2018년 4월 27일)

10) 역사적인 판문점 선언 이행을 위한 군사분야합의서(2018년 9월 19일)

11) 북미 싱가포르 정상회담 공동성명(2018년 6월 12일)

12) 주요 남북합의서

가) 7.4 남북공동성명(1972년 7월 4일)

나) 남북 사이의 화해와 불가침 및 교류 · 협력에 관한 합의서(1991년 12월 13일)

다) 6.15 남북공동선언(2000년 6월 15일)

라) 남북관계발전과 평화번영을 위한 선언(2007년 10월 4일)

마) 한반도 평화와번영 통일을 위한 판문점선언(2018년 4월 27일)

바) 평양공동선언(2018년 9월 18일∼20일 평양)

정전협정전문

http://www.mofa.go.kr/www/brd/m_3984/down.do?brd_id=10909&seq=341009&data_tp=A&file_seq=1

유엔 결의안 82호

http://www.mofa.go.kr/www/brd/m_3874/view.do?seq=311420&srchFr=&srchTo=&srchWord=&srchTp=&multi_itm_seq=0&itm_seq_1=0&itm_seq_2=0&company_cd=&company_nm=

유엔 결의안 83호

http://www.mofa.go.kr/www/brd/m_3874/view.do?seq=310027&srchFr=&srchTo=&srchWord=&srchTp=&multi_itm_seq=0&itm_seq_1=0&itm_seq_2=0&company_cd=&company_nm=&page=35

유엔 결의안 84호

http://www.mofa.go.kr/www/brd/m_3874/view.do?seq=310031&srchFr=&srchTo=&srchWord=&srchTp=&multi_itm_seq=0&itm_seq_1=0&itm_seq_2=0&company_cd=&company_nm=&page=35

육해공 지휘권 이양 서한

http://theme.archives.go.kr/viewer/common/archWebViewer.do?singleData=Y&archiveEventId=0049259188

중공을 한국에서의 침략자로

https://documents-dds-ny.un.org/doc/RESOLUTION/GEN/NR0/744/45/ IMG/NR074445.pdf?OpenElement

휴정조인에 관한 이승만 대통령의 성명

http://syngmanrhee.kr/20/?q=YToyOntzOjEyOiJrZXl3b3JkX3R5cGUiO3M6MzoiYWxsIjtzOjQ6InBhZ2UiO2k6MTt9&bmode=view&idx=289489&t=board

한국휴전에 관한 참전 16개국 공동정책선언

https://www.fordlibrarymuseum.gov/library/document/0337/24468965.pdf

16개국 제네바 공동성명

https://history.state.gov/historicaldocuments/frus1952-54v16/d237

판문점 선언

http://www.mofa.go.kr/www/brd/m_3976/view.do?seq=368007&srchFr=&srchTo=&srchWord=&srchTp=&multi_itm_seq=0&itm_seq_1=0&itm_seq_2=0&company_cd=&company_nm=&page=1

군사분야합의서

https://news.joins.com/article/22985975

싱가포르 회담 공동성명

http://www.mofa.go.kr/www/brd/m_3973/view.do?seq=367939&srchFr=&srchTo=&srchWord=&srchTp=&multi_itm_seq=0&itm_seq_1=0&itm_seq_2=0&company_cd=&company_nm=&page=1

7.4 남북공동성명

http://www.mofa.go.kr/www/brd/m_3984/view.do?seq=341001&srchFr=&srchTo=&srchWord=&srchTp=&multi_itm_seq=0&itm_seq_1=0&itm_seq_2=0&company_cd=&company_nm=&page=3

남북 사이 화해와 불가침 및 교류협력

http://www.mofa.go.kr/www/brd/m_3984/view.do?seq=341000&srchFr=&srchTo=&srchWord=&srchTp=&multi_itm_seq=0&itm_seq_1=0&itm_seq_2=0&company_cd=&company_nm=

6.15 남북공동선언

http://theme.archives.go.kr/next/unikorea/six/six04.do

남북 관계발전과 평화번영

http://www.mofa.go.kr/www/brd/m_3984/view.do?seq=364829&srchFr=&srchTo=&srchWord=&srchTp=&multi_itm_seq=0&itm_seq_1=0&itm_seq_2=0&company_cd=&company_nm=&page=1

평양공동선언

http://www.mofa.go.kr/www/brd/m_3973/view.do?seq=367940&srchFr=&srchTo=&srchWord=&srchTp=&multi_itm_seq=0&itm_seq_1=0&itm_seq_2=0&company_cd=&company_nm=&page=1

Shin, Sang-bum

Military Career

- Senior Member, United Nations Command Military Armistice Commission (UNCMAC)
- Deputy Chief of Staff, ROK-U.S. Combined Forces Command (CFC)
- UC4, Assistant Chief of Staff for Logistics, CFC/United Nations Command (UNC)
- G4, ROK Army HQ
- Chief of Staff, ROK Army Logistics Command
- Commander, ROK 56th Division
- G4, First ROK Army
- Military Advisor, Office of the Minister of National Defense, ROK

Education/ Study

- Korea Military Academy (Class #41), commissioned as 2nd Lieutenant
- Master of Business Management, U.S. Air Force Institute of Technology
- ROK Army College
- GLMP(Global Logistics Management Program) 28기
- The roles of offset organization in the republic of Korea and the republic of China
- Customer-focused perfomance-based logistics support system
- The study of integrated cadet training system
- The study of NANO era defense management

Awards

- Order of National Security Merit (Cheonsu Medal) (2017)
- Presidential Unit Citation (2016)
- Presidential Citation (2008)
- Army Chief of Staff Citation (2007)
- Minister of National Defense Citation (2005)

신상범

군경력

- 유엔군사령부 군사정전위원회 수석대표
- 한 · 미 연합군 사령부 부참모장
- 한 · 미 연합군 및 유엔군사령부 군수참모부 부장
- 육군 본부 군수참모부 부장
- 육군 군수사령부 참모장
- 육군 보병 제56사단 사단장
- 육군 제1야전군사령부 군수처장
- 국방부 장관실 군사보좌관

교육/ 연구활동

- 육군사관학교 졸업 및 41기 육군 소위 임관
- AFIT(Air Force Institute of Technology. 미국 공군대학원) 경영학 석사
- 육군대학교 졸업
- 국제물류관리과정 28기
- 한국과 대만의 절충교역을 위한 조직들의 역할(1997)
- 사용자 중심의 군수지원 성과 관리체계 구축(2005)
- 통합훈련체계 발전 방안 연구(2007)
- NANO 시대의 국방경영에 대한 연구(2016)

상훈

- 보국훈장 천수장 수상(2017)
- 대통령 부대표창 수상(2016)
- 대통령 개인표창 수상(2008)
- 육군참모총장 개인표창 수상(2007)
- 국방부 장관 개인표창 수상(2005)

함께 한 사람들

• 유엔군사령부
- 사령관 : 에이브럼스 육군 대장(General Robert B. Abrams)
브룩스 사령관(Vincent K. Brooks)
- 부사령관 : 에어 육군 중장(Lieutenant General Wayne D. Eyre)
- 참모장 : 질렛 육군 소장(Major General Mark W. Gillette)
미니한 소장(Michael A. Minihan)
- 비서장 : 해밀턴 육군 대령(Colonel Burke R. Hamilton)
- 통역관 : 김현영(Ms. Kim, Hyun Young)

• 유엔사 군사정전위원회
- 수석대표 : 신상범 육군 소장(Major General SHIN, Sang Bum)
- 한국대표 : 권오석 공군 준장(Brigade General KWON, O Seok)
- 미국대표 : 질렛 육군 소장(Major General Mark W. Gillette)
- 영국대표 : 로이드-존스 육군 준장(Brigade General Huw LLOYD-JONES)
- 순환대표 : 안젤레스 필리핀 해군 대령(CAPT Armil ANGELES)

• 유엔사 중립국감독위원회
- 스위스대표 : 고샤 육군 소장(Major General Patrick GAUCHAT)
- 스웨덴대표 : 그랜스타드 해군 소장(RADM Anders GRENSTAD)

• 한미연합군사령부
- 부사령관 : 김병주 육군대장(General Kim, Byung Joo)
최병혁 육군대장(General Choi, Byung Hyuk)
- 보좌진 : 이윤희 중령, 이정환 대령(진), 서준우 중위, 승병철 상사, 최수안 중사

사진 : 제네바 유엔본부, 베른 국제학교, 스웨덴 회담현장, 빈사의 사자상, 폴란드

Photos : The United Nations Office at Geneva, Berne International School, U.S.-North Korea Bilateral Meeting Venue, The Lion of Lucerne, Poland

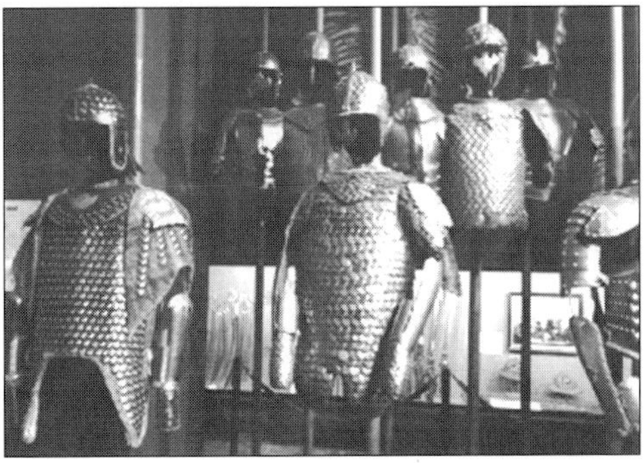

한반도, 정전에서 평화로
The Korean Peninsula, Armistice to Peace

•

지은이 / 신상범
발행인 / 김영란
발행처 / 한누리미디어
디자인 / 지선숙

•

08303, 서울시 구로구 구로중앙로18길 40, 2층(구로동)
전화 / (02)379-4514, 379-4519
Fax / (02)379-4516
E-mail/hannury2003@hanmail.net

•

신고번호 / 제 25100-2016-000025호
신고연월일 / 2016. 4. 11
등록일 / 1993. 11. 4

•

초판발행일 / 2019년 5월 10일

•

•

값 15,000원

•

※잘못된 책은 바꿔드립니다.

※이 책은 유엔군사령부의 보안성 검토를 거쳐 조건부 승인을 받아 출간합니다.

※이 책의 내용은 저자 개인의 견해이며, 정부, 국방부, 연합사, 유엔사의 공식 입장은 아닙니다.

•

ISBN 978-89-7969-799-5 93390